Psilocybin Mushrooms in Their Natural Habitats

I dedicate this book to María Sabina, Tina Wasson, and Kit Scates—three great mycologists and knowledge keepers who have greatly influenced my life.

Psilocybin Mushrooms in Their Natural Habitats

A Guide to the History, Identification, and Use of Psychoactive Fungi

Paul Stamets

TEN SPEED PRESS
California | New York

Contents

Part 2

Identification Guide to Psilocybin Mushrooms

Foreword

Paul Stamets is well known and respected for his pioneering work in mycology, the study of fungi. He has made numerous contributions to the field, including publishing several books on the cultivation of medicinal, psychoactive, and edible mushroom species. He has also collaborated with scientific colleagues to investigate the antiviral and anticancer properties of medicinal mushrooms. His memorable book, *Mycelium Running: How Mushrooms Can Help Save the World* (2005), even lays out a good case for how, through mycorestoration, mushrooms can potentially play a pivotal role in healing our wounded planet, in much the same way that mushroom medicine can restore balance and health to humans.

But it was his encounters with psilocybin mushrooms that originally lured Paul into the multidisciplinary field of mycology, and, in many ways, has remained at the core of his passion for all things myco. He is most widely recognized, one might even say renowned, for his work on the psychedelic, psilocybin-containing species of mushrooms. Paul rightfully deserves credit as one of the most influential scientists and educators—if not the most influential—to bring psychedelic mushrooms to the attention of a wider world. Among his many books, two earlier publications were devoted to psilocybin mushrooms, *Psilocybe Mushrooms and Their Allies* (1978) and *Psilocybin Mushrooms of the World: An Identification Guide* (1996).

Now, with the publication of this latest work, Paul takes his passion to an entirely new level. *Psilocybin Mushrooms in Their Natural Habitats* has it all. It covers the evolution of psilocybin-active fungi, their likely role in co-evolution with humans, their role in ancient and contemporary religious and shamanic practices in both the Old and New Worlds, their emerging importance in medicine, and everything the most devoted psilocybin enthusiasts—or simply interested naturalists—may want to know. It includes chapters on the global distribution of psilocybin-active fungi, where to find them, how to identify them, how to grow them, and (importantly) how to use them safely for mental health benefits and spiritual growth. The second half of the book is an identification field guide to psilocybin-active fungi in six genera, each entry enhanced by excellent color photographs and information on habitats, potency, and other relevant information. The entire book is leavened throughout with stunning color photographs, many taken by the author.

This book is truly Paul's magnum opus, reflecting his nearly fifty years of experience investigating the mysteries of the psilocybian world. The result is the most comprehensive, detailed, and beautiful work on this topic ever published. There have been many books published on the psilocybin-active fungi; none has woven all these divergent threads together in the way that *Psilocybin Mushrooms in Their Natural Habitats* does. It is simply the definitive work on the subject of psilocybin mushrooms—and is likely to remain so for decades.

Dennis McKenna,
president and principal founder of the McKenna Academy of Natural Philosophy, Abbotsford, British Columbia, October 2024

"The Generous Queen" by Autumn Skye

Introduction

The Psilocybin Mushroom Journey Begins

Mushrooms are mysterious: Some can feed you, some can heal you, some can kill you, and some can send you on a life-changing spiritual journey. Psilocybin mushrooms are unique—they could help you live a happier, more fulfilling, more peaceful life and encourage you to be a more responsible Earth citizen. Many people have a hard time describing their psilocybin experiences, which can transcend our everyday vocabulary and spiritual grounding. The ineffable effects can be profound. I propose that psilocybin mushrooms can aid us and, ultimately, our planet—or at least redirect us from our path of self-destruction. Psilocybin mushrooms could help save the world and us as a species!

With *Psilocybin Mushrooms in Their Natural Habitats*, you are embarking on an adventure of self-discovery and immersion into nature. I chose this title to honor Alexander H. Smith and his book, *Mushrooms in Their Natural Habitats.*[1] My professor at The Evergreen State College, Dr. Michael Beug, introduced me to Dr. Smith in the mid 1970s. Both have contributed substantially to the field of mycology, particularly as it relates to the psilocybin mushroom species. These two scientists accepted the sincerity of my interest in psilocybin mushrooms when I was in my early twenties, encouraging me to pursue my studies despite how controversial they were at the time. I remain forever grateful to these two and to all of my mentors.

Psilocybin mushrooms can inspire cosmic visions as well as insight into oneself, the world, and the universe around us. Like recognizing an old friend, once you come to know these beings, they will forever be your fungal allies. And once befriended, they will entice you to photograph, pick, and partner with them to spread their

Psilocybe semilanceata, the Liberty Cap, a potent psilocybin mushroom.

Psilocybe azurescens, one of the most potent psilocybin mushrooms in the world.

Me at 20 years of age, sterilizing grain in my first, albeit unsuccessful, attempt to grow *Psilocybe cubensis*.

spores and mycelium. This book will show you how to embark on an intellectual journey with psilocybin mushrooms and their allies. For many of you, your encounters with psilocybin mushrooms will be sacred experiences. Be careful and respectful of these powerful organisms. They are ancient, having their forms long before we evolved ours.

I have been studying psilocybin mushrooms in their natural habitats since 1974, publishing my first book, *Psilocybe Mushrooms and Their Allies*, in 1979, and my second book, *Psilocybin Mushrooms of the World*, in 1999. This book is my third identification guide, and it reflects my continuing, lifelong dedication to expanding the horizons of knowledge about psilocybin mushrooms, describing many species and growing techniques never before presented.

Mycology is the science of fungi, and mycologists are our fungal knowledge keepers. According to G. C. Ainsworth in his *Introduction to the History of Mycology*, the words *mycology* and *mycologist* debuted in 1836.[2] A very early monograph on fungi, *Rariorum plantarum historia*, dates to 1601, and early mentions of the genus *Psilocybe* include an 1834 report published in the *Transactions of the American Philosophical Society*. Both texts were written in

Latin. There are scant few references on psilocybin mushrooms in these books, in part due to the mushrooms' eclectic and elusive nature. Most psilocybin species do not stand out and are rare to find, with the notable exception of the majestic, dung-loving species like *Psilocybe cubensis*.

Historically, academically trained mycologists with expertise in psilocybin mushrooms were few and far between. As a result, motivated amateurs have contributed substantially to the knowledge base of psilocybin mushrooms. The roles of these passionate citizen scientists, just like in the field of astronomy and so many other areas of study, deepen our understanding of nature. In fact, Charles Darwin was not a classically trained professional biologist. He studied to be a priest, earning a bachelor of arts degree. Yet his theories on speciation and evolution are known as scientific breakthroughs of enormous significance.

"Many of the pioneers in mycology were not professionally employed to work on fungi, but were driven by personal fascinations, and fortunately this remains a key aspect of mycology to this day," observed D. L. Hawksworth in his 2006 article "Mycology and Mycologists" for the 8th International Mycological Congress. "But the skilled enthusiasts ultimately depend on the back-up of professionals, access to reference collections and libraries, and increasingly comprehensive and authoritative databases." Hawksworth even cited an article by R. Watling titled "The Role of the Amateur in Mycology—What Would We Do Without Them!"

And now a new generation of self-taught mycologists has adopted mycology as their profession. The word *amateur* no longer fits. Their passion, experience, intelligence, and on-the-job training bring a fresh perspective to a field that is more invigorated than ever. To this day, relatively few academic mycologists have enough experience to safely guide you to identify, let alone merge with, psilocybin mushrooms.

The current psilocybin awakening is a cross-cultural transformation emerging from the underground. In these times, psilocybin practices are informed by a combination of cultural traditions backed with solid science. With this guide in hand, you join others across the world who steward and share the wisdom of psilocybin mushrooms. Today we are tasked with making choices that affect the future of the social and biological networks that nourish us and our environments. Our collective duty as psilocybin allies is to create networks of collaboration that benefit the Commons while respecting Indigenous traditions, whose histories go back hundreds, if not thousands, of years. Psilocybin mushrooms are also being adopted as sacraments by Indigenous peoples whose lineages do not have documented histories of use. Christian, Judaic, Islamic, Hindu, and Buddhist adherents are currently using psilocybin mushrooms because the common experiences they share strengthen feelings of oneness, forgiveness, kindness, and universal consciousness.[3] Ironically, the very institutions, religions, and cultures that suppressed psilocybin use in the past are increasingly benefitting from the modern use of these sacred mushrooms. Psilocybin mushrooms continue to give us the wisdom and compassion to heal the divisions that keep us apart.

We are all Indigenous to the Earth.

Psilocybin mushrooms unite us.

We are One with the Universe.

***

This guide is designed to help you find psilocybin mushrooms in a variety of habitats across much of the world. Psilocybin mushrooms are diverse—they come in many shapes, sizes, and colors. They grow in fields and forests. They are typically ground dwellers or, in biological lingo, terrestrial. Yet many psilocybin-active species (estimated to exceed 220) share common features that I will describe. By focusing on these traits and learning basic mushroom taxonomy, you are more likely to safely identify psilocybin mushrooms and lessen the risk of accidentally picking poisonous mushrooms. The success of a mushroom field trip depends on sound knowledge, timing, and habitat awareness. Over time, and if you pay careful attention to details, you will be better informed about which species produce psilocybin and which do not. This book will not make you an expert; however, provided you heed its cautionary advice, it will minimize risks and maximize benefits, while leading you step by step up the learning curve.

Mushrooms are miniature pharmaceutical factories, producing hundreds of thousands of molecules. Psilocybin and psilocin are two molecules of particular interest, being the most potent compounds; i.e., ones that produce psychoactive effects. But do psilocybin mushrooms have other possible benefits? Working with scientists, we have found that psilocybin-related tryptamines, like norpsilocin, a metabolite of baeocystin, are potently antiviral *in vitro*. And there may be other active beneficial molecules hidden within them. Although psilocybin mushroom species appear to be safe, some mushroom species that grow among psilocybin mushrooms produce toxic compounds that can be deadly.

Only about 15,000 mushroom species have been identified out of what researchers estimate to be a total of 150,000 possible mushroom species within a larger fungal genome that includes mushrooms, molds, and yeasts, estimated to contain between 2 and 12 million species. More than 2000 new species of fungi are being published

How Many Poisonous, Culinary, and Psilocybin Mushrooms Are There?

Here are some general estimates. Note that the numbers tend to increase the more we explore.

All Types of Fungi:
From 2 to 12 million, and perhaps as many as 20 million species of fungi exist. About 10 percent of fungi are mushroom forming. Only 15,000 to 20,000 mushroom species have been identified out of 150,000 species estimated to exist.[4]

Poisonous Mushrooms:
Approximately 250 species[5]

Edible, Culinary Mushrooms:
Approximately 300 species[6]

Psilocybin Mushrooms:
Approximately 220 species
Psilocybe: 168 species[7]
Panaeolus: 17 species[8]
Gymnopilus: 16 species[9]
Pluteus: 8 species[10]
Inocybe: 6 species[11]
Conocybe/Conocybula: 4 species[12]
Galerina: 1 species[13]

Can a mushroom contain both psilocybin and a toxin? Rarely, but we know of one report (that needs to be reaffirmed): a collection of *Gymnopilus junonius* (= *Gymnopilus spectabilis*) purportedly contains psilocybin and neurotoxic oligoisoprenoids.[14] We will likely find more than one species of mushroom that produces both psilocybin and toxic alkaloids, even amatoxins (bicyclic octapeptides found in deadly Amanitas) within the same fruitbodies. I do not see these compounds as being mutually exclusive. Yet we still do not know why mushrooms produce psychoactive and toxic molecules. Are they incidental artifacts of some other essential metabolic pathways that help mushrooms survive?

each year. These estimates keep expanding as the fungal genome is explored. Each species has a unique chemistry. I see mushrooms as molecular wizards, adept at producing a vast array of secondary metabolites that help the species compete, cooperate, and survive. Many of these metabolites have implications for human and habitat health, and psilocybin is just one stellar example of this.

It turns out that known psilocybin mushrooms, known poisonous mushrooms, and known tasty edible mushrooms are similar in their numbers. Each of these three groups make up less than 2 percent of all characterized mushroom species to date. My greatest concern? On many occasions, I have found psilocybin mushrooms growing side by side with deadly poisonous ones—sometimes close enough to touch (see the photo here and on pages 14 and 171). Picking mushrooms indiscriminately when psilocybin and poisonous mushrooms grow so closely together could be a fatal mistake.

Psilocybe stuntzii, a purple brown–spored, psilocybin-active species, in contact with *Galerina marginata*, a deadly poisonous rusty brown–spored mushroom (also see page 122). Careless collectors could easily comingle these species.

A telltale characteristic that most psilocybin mushrooms share is that they usually bruise bluish. But not all gilled mushrooms that bruise blue are psilocybin species. For instance, the tantalizing bluing stem of *Inosperma calamistratum* (formerly *Inocybe calamistrata*) has not yet been found to contain psilocybin or psilocin. Some *Mycena* species bruise bluish but, upon analysis, no psilocybin or psilocin has been detected (see pages 71 and 120). Please note that the blue color is not uniquely tied to psilocybin or psilocin. Other compounds, many of which have not yet been identified, are at play. When psilocybin mushrooms do bruise bluish, it is, in part, due to psilocybin degrading in a cascade reaction from enzymes into quinoid psilocyl oligomers—compounds similar to indigo (hence the blue color) that are not psychoactive. Yet these compounds could be precursors and theoretically reconverted back into psilocin.[15] The lesson here: Do not underestimate

Psilocybe pelliculosa (above), a psilocybin mushroom, and a species of *Galerina* (below), found adjacent to one another. Some *Galerina* species are deadly poisonous!

Psilocybe cubensis is the most commonly cultivated and used psilocybin mushroom in the world. The mushroom typically bruises bluish when scratched, a telltale sign that this species contains psilocybin and psilocin. However, not all mushrooms that bruise bluish contain psilocybin.

the multidirectional complex biochemistries of psilocybin mushrooms! Chemists are still unraveling their mysteries. Mushrooms are wizards at concocting exotic molecules.

To minimize your risk of misidentification and help you stay alert to seemingly ambiguous taxonomic similarities, I highlight a handful of psilocybin species that co-occur in nature with a poisonous species in the same generic groups. For example, *Conocybula cyanopus* (formerly *Pholiotina cyanopus*) is psilocybin-active; *Pholiotina rugosa* is deadly poisonous. *Galerina steglichii* is psilocybin-active; *Galerina marginata* is deadly poisonous. *Inocybe aeruginascens* is psilocybin-active; *Inocybe geophylla* is poisonous due to its muscarine (a natural plant alkaloid) content that is sometimes fatal to dogs. Other muscarine-containing Inocybes have killed people, although rarely. Several of the more than 1050 known species of brown-spored Inocybes as well as some species of white-spored Mycenas contain muscarine.[16] Not all of these 1400-plus Mycena species have been analyzed, so there are likely many more Inocybes that contain toxins.

The above-named species demand highly refined, expert identification skills. And most experts I know are reluctant to self-experiment with these psilocybin-active species given so many other safe alternatives. We still do not yet know the full array of toxins, if any, that can lurk with psilocybin species within the genera *Conocybe*, *Pholiotina*, *Galerina*, *Gymnopilus*, *Pluteus*, and *Inocybe*. Although these are academically interesting questions, I discourage bioassaying species in these outlier, potentially dangerous genera—at least for now. Given the unknowns here, I strongly encourage you to stay within the safe guardrails of human experience and prioritize the species of *Psilocybe* and *Panaeolus*.

We know which mushrooms are safe to consume from thousands of years of experimentation and more recently from modern analytical techniques. The good news is that science has, by and large, validated what our ancestors discovered through eons of experimentation. It is best to stay within the confines of what has been tested and determined to be safe when it comes to ingestion.

Although the adage in medicine is that the difference between a drug and a toxin is dose, psilocybin is surprisingly nontoxic, even at extremely high doses. One estimate is that you would have to consume nearly 42 pounds (19 kg) of fresh psilocybin mushrooms to reach the LD50 (a lethal dose with which 50 percent of users would die).[17] For a person weighing 154 pounds (about 70 kg), that dose represents more than one-quarter of their body weight! (Please note that, although psilocybin has extremely low toxicity, many other compounds in mushrooms have not yet been identified or tested.) Since mushrooms in general are highly complex in their chemistry—sometimes containing interacting molecules whose mechanisms of actions are still poorly understood—it is wise for most people to choose the psilocybin species tested and proven by humans over hundreds of years: those within the genera *Psilocybe* and *Panaeolus*.

Focusing on psilocybin mushrooms with purplish brown to black spores avoids the risks associated with *Conocybula* (*Conocybe, Pholiotina*), *Galerina, Gymnopilus, Inocybe,* and *Pluteus* species, which have rusty brown to brown spores—not the purplish brown to black spores characteristic of species in the genera *Psilocybe* and *Panaeolus*. The bottom line is to practice an abundance of caution when trying to identify any mushroom you may ingest, because your mistake could kill you or others. Mushrooms are powerful organisms deserving of respect and serious study.

Warning:
Never consume any mushroom nor give them to others to eat unless you are positive of its identification—*and* it is safe.

Part 1

Psilocybin Perspectives

ELEUSIS
Soul Medicine for the Betterment of Well People
Healing Addiction Trauma & Depression
Stanislav Grof
5HT2A
Neurogenesis
Neural Interconnectivity
Albert Hofmann
Timothy Leary
Roland Griffiths
Psilocybin
To fathom Hell or soar angelic just take a pinch of Psychedelic said Humphry Osmond to Aldous Huxley
Socrates
Idealist Philosophy was based on Visionary Experience
Greek Philosophers were
Aztec Codex
Ergot
Persephone
Abducted Underground by Hades
Springing Out

Chapter 1

Evolution and Historical Use of Psilocybin Mushrooms

At this point, mycologists have documented more than 220 species of psilocybin mushrooms across a wide range of genera. Psilocybin production likely evolved around 67 to 65 million years ago. The evolution of the psilocybin-active fungal species resulted from at least four periodic transfers of genes coding for psilocybin (converting tryptophan into psilocybin), which illustrates a recurring dynamic flow of psilocybin-encoding genes that has led to the many psilocybin-active species, spread over at least nine genera, currently recognized today: *Conocybe, Conocybula, Galerina, Gymnopilus, Inocybe, Panaeolus, Pluteus, Psilocybe,* and the non-mushroom forming cicada pathogen in the genus *Massospora.*

We do not yet know what the first psilocybin species was. We do know, however, that psilocybin genes have transferred from one mushroom species to others. Psilocybin was likely first expressed in wood-rotting species

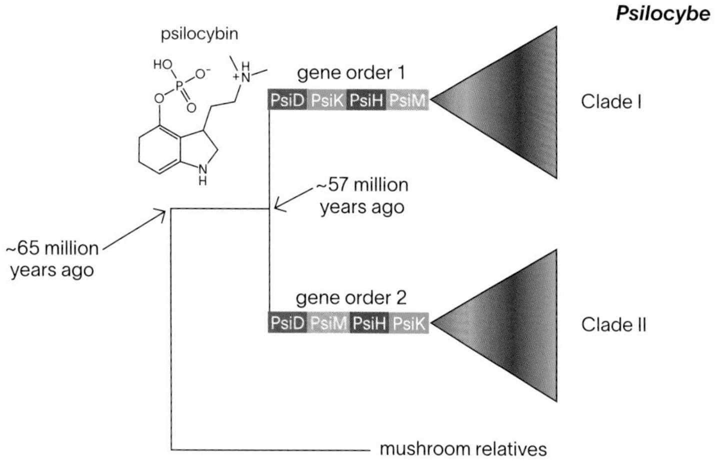

Evolutionary phylogeny or "gene tree" showing estimated origins of psilocybin in Psilocybe. Clade I and Clade II represent the *Psilocybe* species with the same gene order patterns within their psilocybin-producing gene clusters. Image credit: Bryn Dentinger.

"Albert Hofmann and the New Eleusis" by Alex Grey

A *Psilocybe cubensis* related species growing on elephant dung.

Psilocybe cubensis growing on cow dung.

and later appeared in dung dwellers. There are many possible ways for psilocybin genes to have spread. Bradshaw et al. noted that an ecological shift to dung species occurred independently on at least two occasions, citing four or five possible horizontal gene transfers flowing to other mushrooms from 40 to 9 million years ago.[1]

Interestingly, the arrival of the gene cluster that codes for psilocybin is coincident with the Cretaceous-Paleogene extinction event approximately 66 million years ago.[2,3] After that extinction event, which was triggered by an asteroid impact, more than 75 percent of species died off. Megafauna herbivores, such as the early ancestors of elephants and mastodons, began rapidly differentiating thereafter. Multiple lineages led to the dung-producing, grass-grazing megafauna. The lineage of these animals can be traced to 50 to 25 million years ago. Psilocybin mushrooms were co-evolving during this same period. Our modern-day cattle trace their origin to 12 million years ago. I suspect that psilocybin mushrooms may have made the jump from megafauna, such as elephants or hippopotamuses, to bovines during this same time period.

Two of many possible vectors of gene transference are rotting bacteria and anastomoses (mycelial fusion or proximate contact), aided by insects and birds, especially egrets, that likely spread psilocybin mushrooms to new habitats. Several pastoral mushroom species may have benefitted, including the majestic *Psilocybe cubensis* and the newly named *Psilocybe ochraceocentrata*. *Psilocybe cubensis* is now the most popularly grown and consumed of all psilocybin mushroom species. These species can co-occur on the same pile of dung with likewise potent species in the *Panaeolus cyanescens* taxonomic cluster. One working hypothesis is that psilocybin genes first appeared in dung-dwelling species of the *Panaeolus cyanescens* group and jumped to dung-dwelling *Psilocybe* species, like *Psilocybe cubensis* or *Psilocybe ochraceocentrata*. Many avenues have led psilocybin to move from one species to another. And they are likely continuing to spread, with more psilocybin-active species evolving and yet to be discovered.

The pastoral preferences of the *Psilocybe cubensis* group allowed it to spread among species of megafaunal herbivores as forests were cleared and replaced with grasslands for cows, sheep, and horses. Since the pantropical *Psilocybe cubensis*–related taxa continue to be found on elephant dung, its jump to the dung of wild ox and the descendent cows are likely a broadening of habitat—and a more recent occurrence. An interesting and plausible theory is that egrets helped spread spores of *Psilocybe cubensis*–related species after forming a mutualistic relationship with cattle.[4] Egrets feast on swarming insects, consuming the insects that take flight as cattle walk and disturb the habitat underfoot. The egret hypothesis is especially interesting to me, as egrets sit upon many megafaunal herbivores. Many insects thrive in not only the dung, but also inside most of the mushrooms growing on the dung. For many mushrooms, insects are a primary vector of spore and mycelium distribution. Since egrets and dung-dwelling mushrooms both prefer ponds, marshes, shores, and swamplands, they have frequent encounters. As migratory birds, egrets can then extend the mushrooms' distributions still further. Many other birds, including crows and ravens, can be ravenous consumers of mushrooms.

Why would a mushroom species produce psilocybin? What evolutionary advantage would psilocybin confer? Since psilocybin mushrooms predate humans and all of our extinct bipedal ancestors—many had their forms long before our primate ancestors evolved—a biological explanation independent of human development seems likely.[5] Some scientists have proposed that psilocybin is a potential insecticide, preventing insect predation through disorientation or loss of appetite—abating insects' mycovorous tendencies to eat the fruitbodies.[6] However, I find fly larvae inside many maturing psilocybin mushrooms, especially pastoral species. What I have noticed repeatedly is that another group of invertebrates—slugs and snails (terrestrial members of the phylum Mollusca)—avoid psilocybin-active mushrooms in favor of psilocybin-inactive species. I did a simple choice test with invasive European black slugs (*Arion* spp.) placed into a cardboard box with the wavy-capped *Psilocybe cyanescens* and the Garden Giant, *Stropharia rugoso-annulata*, at opposite ends. The slugs swarmed over and ate the gills and flesh of the Garden Giants but did not consume the Wavy Caps. When given only *Psilocybe cyanescens* mushrooms, the slugs seemed disinterested and left them unharmed. Of course, two tests are not conclusive and this theory needs to be proven, but what we witnessed was impressive.

Like other mushroom species, psilocybin mushrooms attract sciarid, phorid, *Drosophila*, and other flies that spread spores to new habitats. Slugs and snails, on the other hand, devour the flesh of the emerging fruitbodies and consume down to the spore-producing gill layers, abating spore release. This observation supports the theory that psilocybin could prevent gastropod (slug and snail) predation. But, if this is the case, why don't all mushrooms produce psilocybin? That psilocybin mushrooms also use humans to spread their spores may be a later development that helped ensure their survival, as has been the case with many plants that humans have cultivated—from apples to marijuana.[7] Psilocybin mushrooms inspire us to consider them as sacred allies to be protected and cultivated. By accident or design, they have achieved a new vector for panspermia—hitchhiking with us as we migrate and colonize new habitats—even, plausibly, as we explore space. I propose that astronauts will need psilocybin mushrooms for psychological (and ecological) benefit to overcome the loneliness during our extraterrestrial explorations to other planets—and to stay on mission.[8] We are now bonded with psilocybin species forever.

The Stoned Ape Theory

Terence and Dennis McKenna popularized the stoned ape theory in the 1990s. The McKennas proposed that as climates changed, our bipedal ancestors had to adapt and innovate to survive in new habitats. The brothers credited the consumption of *Psilocybe cubensis* with the sudden evolution of brain enlargement that resulted in *Homo sapiens* approximately 200,000–300,000 years ago because high doses of psilocybin sparked neurogenesis. Now this hypothesis, once widely ridiculed, does not seem so far-fetched. Although at the time they proposed this idea, it was just a hypothesis, given the evidence we have today, I think it now qualifies as a theory worthy of serious consideration.

When our bipedal ancestors left the forests to track megafauna in the grasslands, they would have likely encountered *Psilocybe cubensis* (or the closely related *Psilocybe ochraceocentrata*) growing from the dung of cattle, zebras, antelopes, elephants, hippopotamuses, rhinoceroses, and possibly other African megafaunal herbivores. Those who have collected *Psilocybe cubensis* (aka Golden Top) can see their fruitbodies from hundreds of feet away—they are that obvious. When mushrooms grow on dung and sporulate, flies are attracted and lay eggs. Aging *Psilocybe cubensis* is often swarming with larvae and, since many primates eat maggots, this would have been seen as a potential source of nutrition. Twenty-two species of primates—twenty-three when you count humans—consume mushrooms for food.[9] In fact, the Goeldi monkey of the Bolivian Amazon consumes approximately twelve times its body weight per year in mushrooms. Many primates know which ones are edible and which ones to avoid. Did our primate ancestors devour mushrooms then as they do today? They undoubtedly encountered these species, but did consuming psilocybin mushrooms stimulate enlargement of the primate brain?

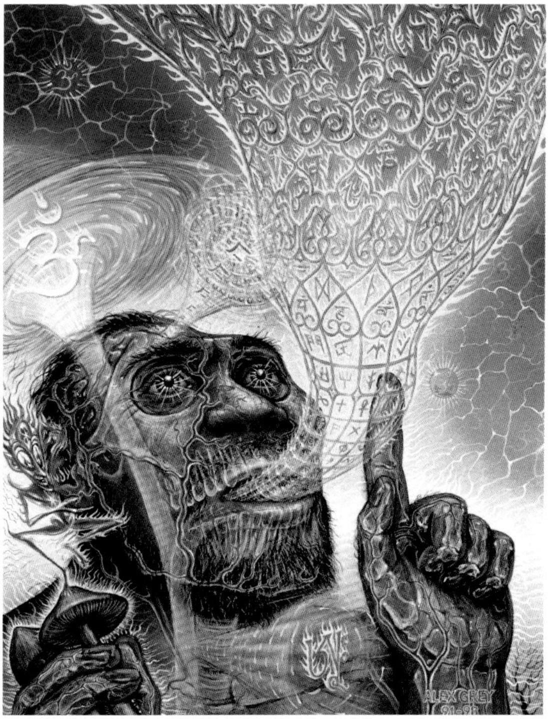

Alex Grey's depiction of the stoned ape theory, which proposes that the frequent consumption of psilocybin mushrooms increased cognitive function and brain matter and spurred the evolutionary arrival of *Homo sapiens*.

Our ancestors likely consumed psilocybin mushrooms many times over millions of years, making psilocybin a possible epigenetic neurostimulant, meaning new genetic factors for neurogenesis emerged. Through the mechanism of epigenetics, external influences (e.g., psilocybin and climate change) might influence gene expression to increase survival. Family and fellow gatherers could self-dose for revelatory and entertaining communal experiences (trips), in addition to having a nutritious meal. Would such encounters benefit the evolution of our primate ancestors' brains and confer new neural connections? Scientists continue to compete by postulating theories on how and why the cranial volume of our early ancestors expanded by a factor of two or three in less than 200,000 years to reach the size and complexity of our brains today. Our species *Homo sapiens* evolved relatively recently. The McKennas speculated that by repeatedly ingesting psilocybin mushrooms new neurons formed, enlarging brain capacity, which allowed our apelike ancestors to develop new skill sets for survival.

I think this untested hypothesis should be elevated to a testable theory. It seems even more plausible since new research has shown that psilocybin and other closely related tryptamines increase nerve growth in pluripotent (progenitor) human stem cells.[10] We also now have evidence that psilocin binds with neuroreceptors that stimulate neurogenesis, neurogeneration, and neuroplasticity (for more, see page 109). In their time and to this day, however, the McKennas' hypothesis was roundly criticized by scientists in various fields as ridiculous, merely a stoner's dream. Naive critics venture "expert" opinions limited by the lack of first-hand psychedelic education.

Cultural Uses of Psilocybin Mushrooms in the Old and New Worlds

Although we lack much written documentation of psilocybin mushroom use in the Old World compared to the New World, threads of ancestral knowledge conveyed by oral traditions have survived. For instance, Basotho healers, called *lingaka* (doctors), from the Kingdom of Lesotho in southern Africa, have a long oral tradition of using *Psilocybe maluti* (see page 169), called *koae-ea-lekhoaba*, steeping the mushrooms in hot water along with the hallucinogenic plant *Boophone disticha*, which is known as *seipone* or *leshoma*.[11,12] Giorgio Samorini also reported on the possible use of divinatory mushrooms in North Africa and the Ivory Coast.[13,14] I suspect these are but a few of many examples of psilocybin use in prehistory. Many more have yet to come to the attention of present-day scientists. I am convinced that Indigenous wisdom is, in its essence, a form of science. Both schools of science, ancestral and current, are nature based.

We do have intriguing physical evidence that may bear witness to ancient psilocybin use in the Old World. We see artifacts of probable depictions of psilocybin mushrooms in archaeological records by paleolithic and neolithic artists. These images are, of course, open to interpretation, but the frequency with which certain mushrooms appear is compelling. I rely not only on what others have written, but also on discoveries not yet reported that support the historic body of evidence of probable psilocybin use in the Old World. I encourage serious academic study of the use of psilocybin mushrooms—past and present—across the planet. Some regions of the world are underrepresented due to cultural oppression, as psilocybin mushrooms unshackle one's consciousness from the control of dominant dogma and ideology. Another reason for underrepresentation in regions that may have had psilocybin histories is climate change. Mushrooms that were once common along lush riparian zones have declined or become extinct from desertification. As mushrooms become scarce, Indigenous use and knowledge also decline and mycological wisdom can be lost.

Admittedly mycologists have a mycocentric point of view, but we also contribute eclectic knowledge that many historians should consider before outright discrediting new theories. Only recently has the study of mycology been elevated to a new level of scientific credibility, having

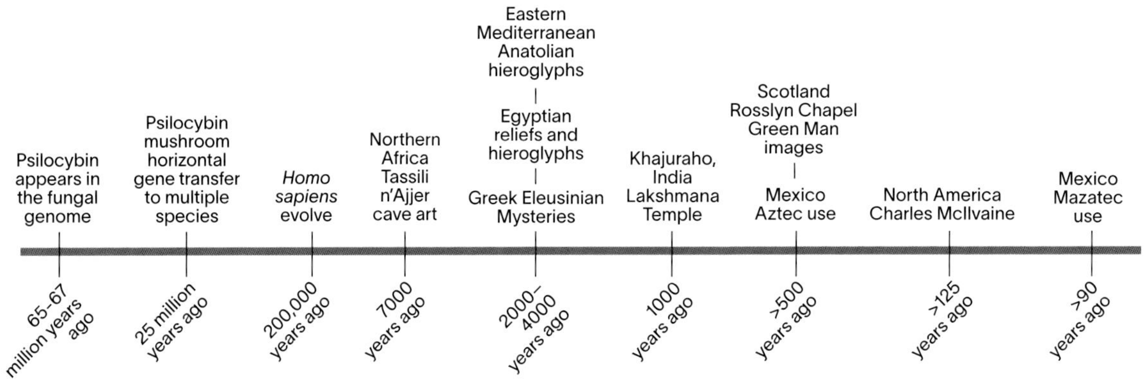

Putative historical timeline of some events representing possible and documented psilocybin mushrooms in human culture.

been previously relegated to a science of little consequence. I am open to counter opinions and respectful dialogue on this issue, as it is fundamental to not only the evolution of civilization but also to our spiritual views of the universe.

Old World Psilocybin Mushroom Use: The Case for Probability

The Tassili n'Ajjer cave art of the Bee-Mushroom Man is more than 7000 years old. Several depictions are inscribed on cave walls on the Tassili n'Ajjer plateau of present-day southeastern Algeria, Libya, Niger, and Mali. *Tassili n'Ajjer* translates as the "plateau of running rivers." Millennia earlier, the area had a moister climate than today's now-dominant, arid Sahara. Long-horned cattle frequented this habitat, and presumably *Psilocybe cubensis* or *Psilocybe ochraceocentrata* could have been common. The figure of the so-called Bee-Mushroom Man is intriguing not only because of the numerous mushrooms depicted, but also because the taxonomically accurate *shapes* of the mushrooms resemble many psilocybin varieties. In Mesoamerica, Indigenous people realized that placing mushrooms in honey was an excellent way to preserve these very perishable species, so the association of bees, honey, and mushrooms in this cave art makes sense.

Potentially the earliest hints of psilocybin mushroom use can be found in the Tassili n'Ajjer caves in Africa; the Selva Pascuala cave near Villar del Humo in Spain; Stonehenge in England; the Telesterion at Eleusis in Greece; the Dendera Temple of Hathor, Debod, and Isis (and many others) in Egypt; the Lakshmana temple in Khajuraho, India; and the Rosslyn Chapel in Scotland.

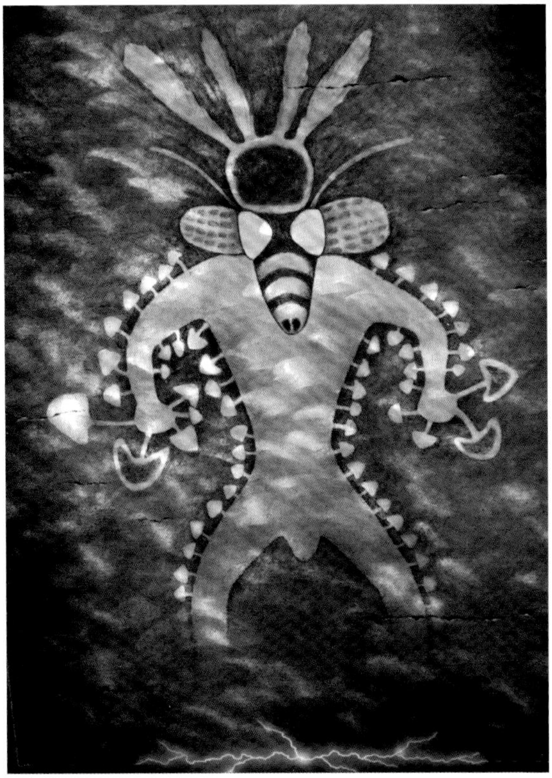

The Bee-Mushroom Man recreated on copper by artist Jonathan Meader.

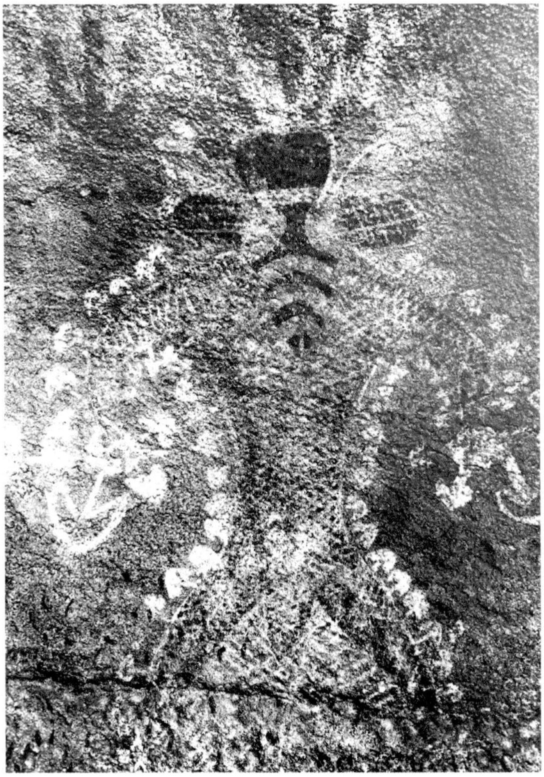

The Bee-Mushroom Man cave art is estimated to be 7000 years old. From *Merveilles du Tassili N'Ajjer* (*Wonders of Tassili N'Ajjer*) © 1962 by Jean-Dominique Lajoux, Arthaud, Paris.

The lack of acknowledgment by archaeologists for mushrooms being portrayed in what is colloquially called the Tassili Bee-Mushroom Man is just one of many examples of what I call mycological myopia—a failure to recognize representations of mushrooms for what they are. In his book *The Rock Paintings of Tassili*,[15] French explorer Jean-Dominique Lajoux failed to mention mushrooms even though the pictographs abound with running figures adorned with mushroom-like heads carrying cuspidate mushrooms in hand. The drawings of the mushrooms are morphologically consistent and even match the phonotypes of many psilocybin mushrooms.

To quote Lajoux, "This painting has the highest degree of symbolism... combined in a pattern whose meaning escapes us.... A crudely drawn white bovid... is mixed up with these figurations and plays an effective part in the total composition." He also remarks on mushroom-shaped

The Running People pictograph shows a taxonomically accurate *Psilocybe* mushroom phenotype and directly connects (with dots possibly representing spores or ideas) the mushrooms to the human-like heads.

images without suggesting they represent mushrooms, noting that "These masked figures are the only ones on this site. They are *olive green*, outlined in *purple*." Psilocybin mushrooms are often greenish to purple because of their psilocin content. This is yet another element that convinces me that the images portrayed are more likely psilocybin mushrooms than not.

Psilocybe cubensis, Psilocybe ochraceocentrata, and *Psilocybe fimetaria* all grow on bovid (cow) dung, often have pointed caps (cuspidate), and turn olive green to bluish purple when picked. The bee-masked head and numerous mushrooms adorning the humanoid figure underscore the close association of mushrooms, cows, bees, and, by extension, honey. To me and many other mycologists, this appears to be a psilocybin pictorial field guide and preservation ideogram from paleolithic times.

The 6000- to 8000-year-old cave mural called Selva Pascuala, near the town of Villar del Humo, Spain, shows a long-horned aurochs (*Bos primigenius*, the Holocene ancestor of modern domesticated cattle) with about a dozen conic mushrooms sprouting from the ground nearby. Many psilocybin mushrooms feature a conic cap that narrows at its top and have a stem longer than the width of its cap. This cave art is unlike porcini or button mushrooms, which feature rounded caps that are wider than the stems are tall. Several researchers have suggested that the depicted species could be *Psilocybe hispanica*, a species likely to be synonymous with *Psilocybe fimetaria*.

Until 400 CE, the Greeks engaged in a nearly 2000-year-old tradition known as the Eleusinian Mysteries, an autumn ritual of death and rebirth celebrating the goddess Demeter and her daughter Persephone. Greek citizens by the hundreds, including enslaved people, partook in a sacred ceremony during which they walked many miles to a great telesterion (temple), whereupon, it is theorized, they ingested a potent psychedelic elixir, a *kykeon*, in a group setting in the Temple of Eleusis. In their classic 1978 book, *The Road to Eleusis: Unveiling the Secret of the Mysteries*, R. Gordon Wasson, Albert Hofmann, and Carl Ruck argued that the Eleusinian Mysteries incorporated fungi—likely a detoxified extract from the ergot fungus (*Claviceps purpurea*), a natural source of lysergic acid alkaloids (related to LSD)[16] that grow on wheat, rye, darnel, oats, barley, and other grasses. The authors also speculated that the name Demeter is etymologically similar to *erysibe*, or ergot. In his 2020 book *The Immortality Key*, Brian Muraresku reached a similar conclusion based not only on the literature as analyzed by Wasson, Hofmann, and Ruck, but also on biochemical analyses of ancient chalices used in similar rites that were found to contain ergot residues.[17]

Psilocybin mushroom meads (fermented honey beverages) were likely consumed in Europe for centuries. Placing psilocybin in honey is a popular practice that continues today with the use of Shape Shifter Honey, psilocybin-infused honey, a popular addition to teas or hot water at such festivals as the Telluride Mushroom Festival, Burning Man, and the Glastonbury Festival. That this practice is so widespread today is a testimonial of shared knowledge that extends back thousands of years, tracing from multiple cultures.

The grassland habitats of the Pyrenees Mountains of Spain and France host *Psilocybe fimetaria*, a psilocybin mushroom that grows on horse and cow dung.

Cave art in Spain, estimated to be 6000 to 8000 years old, depicts a long-horned cow (aurochs) with cuspidate psilocybin-like mushrooms nearby. That mushrooms are depicted suggest they were important to our prehistoric ancestors. That a cow is nearby suggests the two are interrelated.[18]

Moreover, since ergot grows with Mediterranean grasses and psilocybin mushrooms grow in the same grasslands in the highlands above Eleusis, I would not be surprised if both were employed in the ceremonies. Pairing two psychoactive fungi would be a powerful psychoactive concoction. Although there does not yet exist concrete evidence that the Greeks used psilocybin mushrooms during the Eleusinian

This stone relief features the goddess Demeter giving her daughter Persephone a mushroom before Persephone enters the autumnal underworld, where she would slumber through the winter, only to reemerge in the spring.

Mysteries, we do have this intriguing relief from 2400 years ago that directly connects Demeter and Persephone with mushrooms as Persephone enters the Underworld, implying that the mushroom's narcotic effects helped her in the journey.

The Roman Emperor Augustus and philosophers such as Socrates, Cicero, and Euripides also partook of these elixirs, which were made available to most Greek citizens, although the costs were equivalent to several months' wages of the common people. None could speak of this practice for fear of being caught, the penalty for which was death.[19] In his self-published, must-read epic treatise *Pharmacotheon*,[20] the brilliant Jonathan Ott elaborated on the Eleusinian Mysteries, speculating that one of the reasons Socrates was put to death was because he betrayed the secret of the Eleusinian Mysteries. I think the priests of Eleusis were, in effect, skilled mycologists. They had centuries to perfect these elixirs. We do not know how the mycologist-priests might have created Eleusis's potion, one that could excite consciousness without causing gangrene, a condition that can be caused by eating ergot-moldy grain-based foods. In an

Lakshmi, the goddess of prosperity, is associated with the lotus flower. Here, two hands present her with cuspidate forms that appear to me, as a mycologist, to match the iconic phenotype of *Psilocybe cubensis.*

Notably similar in form to the mushrooms in the Lakshmi relief, this is a grouping of *Psilocybe cubensis*, which grows from the dung of cattle, elephants, and other megafauna herbivores. Cows have been considered sacred in India for thousands of years. Why?

interesting review of the making of Egyptian beer, Tom Riedlinger proposed one method for creating such a potion.[21] I think ergotism would more likely occur from repeatedly consuming ergot-contaminated grains, not from a once-in-a-lifetime ingestion. Despite our advances in pharmacology, the Eleusinian Mysteries continue to be a mystery to this day.

We also have tantalizing evidence of sacred mushroom use in India. Although Wasson proposed that the intoxicant, Soma of India, was *Amanita muscaria*,[22,23] I remain skeptical. More enticing to me is that *Psilocybe cubensis* was used for spiritual purposes. A relief from the Lakshmana Temple, built by the ruler Yashovarman about 1000 years ago in Khajuraho,

Bread made from ergot-contaminated grains were also likely the source of the well-documented Saint Anthony's Fire in medieval Europe, causing temporary "madness" and a burning sensation in the arms and legs.[24] The dire symptoms and alteration of consciousness supposedly evidenced witchery. Hundreds of thousands of people perished during the inquisitions that were inflicted upon wisdom keepers and loosely defined heretics accused of knowing how to induce altered states of consciousness. Practices of imbibing concoctions that change one's worldview and liberate consciousness have been systematically suppressed, since they were seen as threats to the established order. A more recent example is President Richard Nixon's War on Drugs, which outlawed psilocybin and other psychedelics in 1971, targeting a large contingent of his political opponents, particularly minorities and the young.

This Libertas Americana medal from 1783 celebrated the independence of the United States from Great Britain on July 4, 1776. This iconic image can be traced to the liberty cap worn by emancipated Roman slaves, which was later resurrected to champion the freedom of citizens from authoritarian rule after the French Revolution in 1789. It is also known as the Phrygian cap. In mycology, the common name Liberty Cap became intimately associated with *Psilocybe semilanceata* due to its morphological similarity. In more modern times, this species of psilocybin mushroom induces the liberation of consciousness.[25]

India, depicts phenotypically accurate forms resembling *Psilocybe cubensis*, which can still be collected sprouting from cow dung nearby.[26] Shiva, the Lord of Intoxicants, has a blue throat, as the mushrooms bruise blue, and cows are culturally protected and revered in India, perhaps because they are the ultimate purveyors of the psilocybin mushrooms that emerge from their dung. Both of these elements support my not-so-far-fetched hypothesis. Curiously, legend states that the Buddha died (c. 483 BCE) at the age of 80 from eating a type of mushroom,[27] which some propose was his final act of enlightenment.

We also have mushroomic reliefs carved into the Rosslyn Chapel, built in the fifteenth century in Scotland; the reliefs celebrate the diversity of religions, from pagan to Catholic to Celtic to Nordic peoples. When I visited Rosslyn Chapel in 2010, I was awestruck by the number of mushroom-looking carvings etched into the blocks from which this temple was constructed. Moreover, the Rosslyn Chapel is surrounded by pastures grazed by sheep, an ideal habitat for the Liberty Cap, *Psilocybe semilanceata*. Anthropologist Jerry Brown, PhD, and Julie Brown also explored this temple and speculated on the use of psychedelics in early Christianity.[28]

Although Stonehenge, located in the Salisbury Plains of England, predates the Druids according to some experts, I was amazed to see sheep-grazing pastures that are ideal for growing

One of the 115 box reliefs of Green Man adorning the inside of the fifteenth-century Rosslyn Chapel in Scotland. Green Man was a mythological figure representing the need to connect to nature, return to earth. These motifs have long appeared in both pagan (including Celtic) and Christian cultures. Many have cuspidate mushroom-like forms emanating from the mouths. Interestingly, from my count, there are ten times more representations of Green Man than of Jesus Christ. The conic cap, enrolled margin, and long stem are features typical of *Psilocybe semilanceata*, a species growing in the sheep pastures adjacent to this temple.

The Liberty Cap mushroom, *Psilocybe semilanceata*. The name liberty cap is now uttered more in its association with a mushroom than its historical reference to a Phrygian hat.

From more than 2000 years ago (c. 50 to 25 BCE), this ancient hearth was found in Welwyn, England, about 100 miles from Stonehenge. Sprouting forms resembling Liberty Caps, *Psilocybe semilanceata*, a potent psilocybin mushroom that still grows in the local pastures that have been grazed by cows and sheep for centuries. Housed at the British Museum, collection #1911,1208.2.

Liberty Caps (*Psilocybe semilanceata*) directly around this ancient, majestic ceremonial site. Whether Druids—the high-ranking priests of the Celts—of that region used psilocybin mushrooms long ago is debatable. What is not debatable is that the Druids currently use psilocybin mushrooms and consider Stonehenge a sacred temple for psilocybin mushroom journeying. This is yet another example of a connection that may never be proven—or disproven.[29] The circumstances of psilocybin mushrooms resident in the ecosystems hosting these spiritual monuments are fuel for speculation of a past long ago hidden from modern eyes. The passage of time has obscured our ability to know, but not to hypothesize.

Umit Sayin, a Turkish physician, made the case that the ancient Anatolian cultures of Asia Minor, which spans present-day Turkey and Syria, also used psilocybin mushrooms as evidenced by reliefs and statues from 1500 to 500 BCE.[30] An Anatolian relic celebrating the goddess Kubaba has mushroom-like hieroglyphs carved at its base (see opposite page).[31] Although some experts are convinced of the ancient use of psilocybin mushrooms in this region, others are not.

Did the ancient Egyptians use psilocybin mushrooms? It seems likely, as Egyptian culture is rich in symbology with deep spirituality and ties to nature. Yet the ethnomycological traditions of the past depend on our understanding of ancient practices—a challenge since these practices were protected and cloaked by veils of secrecy. Considering the lack of written evidence, we must take into consideration the role of ancient art within the context of the ecosystems that prevailed during those times. If archaeological evidence is unclear, then paleoecological analyses may provide some potential explanations.

The same forces of desertification that drastically altered the region of the Bee-Mushroom Man's Tassili n'Ajjer plateau affected much of North Africa and the Mediterranean region, including the Nile Basin of present-day Egypt. As aquifers dried up, the Nile drew more people for navigation and agriculture. Although mushrooms are scarce in this desert region today, they were clearly part of Egyptian history and culture.

Prized by ancient Egyptians, mushrooms were described as "without leaves, without buds, without flowers: yet they form fruit; as a food, as a tonic, as a medicine: the entire creation is precious."[32,33] The ancient Egyptians considered mushrooms to be a gift from the god Osiris, the god of the afterlife and the underworld. In the Papyrus of Ani, also known as the Egyptian Book of the Dead and dated to 375–275 BCE, mushrooms were called the "flesh of the gods." Interestingly, the Aztecs called some psilocybin mushrooms *teonanácatl*, which translates identically as "the flesh of the gods" (see page 36).[34,35] Across cultures, across continents, and across millennia, mushrooms evoke the divine from those who experience them.

Statue of the goddess Kubaba from the ninth century BCE, currently at the British Museum in London. Note the mushroom-shaped hieroglyphs.

Mycologically naive archaeologists are understandably unlikely to see mushrooms in ancient records and imagery. Finding the DNA of *Psilocybe cubensis* or *Psilocybe ochraceocentrata* in the vases or soils around the Egyptian temples that depicted mushrooms would advance this theory. As technology becomes more sophisticated, many secrets of the past are being revealed—this could yet be one of them.

Berlant,[36] Froese et al.,[37] and Abdel-Azeem et al.[38,39] all proposed that the ancient Egyptians could have used a species such as *Psilocybe cubensis* based on the morphological similarities of hieroglyphs depicting mushroom shapes. In 2013, the legendary Baba Ahati Kilindi Iyi, known for his high-dose use of psilocybin mushrooms, also suggested that Egyptian hieroglyphs depicted *Psilocybe cubensis*.[40] This hypothesis makes sense to me on multiple levels.

In mycology, we look for indicator species, a species associated with the ecosystems that support the mushrooms we seek. In ancient Egypt, the blue lotus was considered sacred and prominent in Egyptian spirituality, frequently portrayed in hieroglyphs and reliefs on the walls of many temples. The blue lotus (*Nymphaea nouchali* var. *caerulea*), a water lily, grows in ponds. Ponds attract cattle seeking water. Where the blue lotus flourishes in ponds with cattle assembling nearby, the coprophilic *Psilocybe cubensis* species group would also likely thrive. Foragers collecting the blue lotus flowers for its edible and

Mushroom forms phenotypically similar to *Psilocybe cubensis* and *Psilocybe ochraceocentrata* appear to be growing out of a vessel held by Hathor, goddess of beauty, who is often represented in her animal form as a cow (30 BCE to 14 AD). See also the image on page 28, a classic form of *Psilocybe cubensis*.

Notice the blue lotus floret with mushroom-like forms together at the feet of the goddess Hathor.

medicinal bulbs would likely encounter *Psilocybe cubensis*-like species fruiting from the dung. I believe the early Egyptians likely knew of these golden-colored, bluing psilocybin mushrooms because they and the blue lotus share the same wetland ecosystem. Portrayals of the two in proximity on certain temple walls show that the Egyptians knew they are intimately associated.

The ancient Egyptians recorded this in carvings as a message for eternity whose meaning has become obscured, as both the blue lotus and mushrooms have become scarce in that region. The average rainfall today is less than 0.04 inches (1 mm) annually. Most notably, the seasonal flooding of the Nile thousands of years ago brought much needed water, moisture, and nutrients, allowing this region to gain the reputation as a breadbasket agricultural delta. It is drastically different today, as droughts impact water resources in neighboring Sudan and Ethiopia, sources of most of the Nile's water. Pollen analysis shows a radical shift in that region to a more arid ecosystem in 1250 to 1100 BCE.[41] Climate change fundamentally changed the course of Egyptian culture.

In 2016, Egyptian mycologist Ahmed Abdel-Azeem proposed that the Dendera Temple wall relief of the goddess Hathor holding a mushroom basket represented the classic forms of the *Psilocybe cubensis* group. In May 2023, we visited nine temples along the Nile River, including the Dendera Temple. On the walls directly below Hathor, we found mushroom-shaped wall reliefs, often together with the blue lotus and the falcon, both accepted symbols of Egyptian spirituality. Egyptians considered the blue lotus, a sacred flower that opens in the day and closes at night, to symbolize life, death, the afterlife, rebirth, and enlightenment. The falcon represented the sun god, the living king of Egypt. That the shapes of a falcon, a blue lotus, and a mushroom can be found repeatedly together underscores their synergistic association.

A mushroom-looking wall relief at the Temple of Edfu (237 to 57 BCE). To many mycologists, this is a taxonomically accurate depiction of a mushroom very similar to many *Psilocybe* species.

For us mushroom cultivators, the vessel held by Hathor makes sense as a fruiting chamber. If you move wild dung piles fruiting with *Psilocybe cubensis* into a vase like the one portrayed on page 32, another flush of mushrooms is likely to grow out of the narrow openings within 1 to 3 weeks. Such a vessel would not only retain moisture, but also eliminate the damaging effects of sunlight and day to night temperature extremes and allow carbon dioxide to escape through the open ports, while allowing exposure to oxygen so the mycelium could breathe. Water could be easily added and slowly evaporate—helping the mycelium to respire. *Psilocybe* mushrooms would form elongated stems as the mushrooms sought light from the moist dark vessel's chamber so they could sporulate in the open air. And once they emerged, the mushrooms would curve up against gravity as depicted. Mushroom cultivators commonly use small openings to produce focused fruitings with many species. If mycologically astute ancient Egyptians had discovered that the stem butts regrow after you cut them off (see page 78), the likelihood of success would be increased if they put them back into the vessel. This is yet another aspect that most Egyptologists who are not skilled in the field of mycology would not know.

Falcons sitting upon mushrooms with emerging blue lotus. Falcons represented the god Horus, the sun god and king of Egypt. That the three are represented together is significant.

Are some of these hieroglyphs and carvings of psilocybin mushrooms? Or are they all just the fanciful imaginations of those with a bias who think psilocybin mushrooms are central to the emergence of modern civilizations? Or are the skeptics biased due to their naivete? Some Egyptologists are at a disadvantage, as they lack knowledge of the prevalent ecosystem thousands of years ago that harbored the cultures they are struggling to understand. For most, mushrooms are outside of their ken as they are now scarce in a bioregion that has undergone dramatic desertification. From a broader perspective, I have learned not to underestimate Indigenous peoples' ability to know intimately the organisms in the ecosystem that generations before them have tested.

The methods I have used for growing psilocybin mushrooms by transplanting mycelium have no technological constraints. These methods could have been practiced thousands of years ago and are being practiced by many today. What is undeniable is that psilocybin mushrooms are visionary to those using them across the world, from the ancient past to this day, invoking deep reverence, spirituality, inspiration, insight, and artistic creativity; benefitting mental health; and helping us become more civil. Artists have codified these relationships for eons. Aficionados want to propagate them.

Recently, I met a new generation of Egyptians in Cairo who are rediscovering or "re-indigenizing" this lost tradition by ingesting elixirs combining blue lotus extracts with the blue-staining *Psilocybe cubensis*. This combination is now undergoing an international resurgence—perhaps a re-remembering of a cultural heritage echoing from the ancestral practices long ago.

As with some conventional Egyptologists, the French ethnologist Jean-Dominique Lajoux could not recognize the mushrooms that were depicted before him because he simply did not know what he was seeing. This is by no means surprising—a desert-focused archaeologist would not recognize portrayals of mushrooms. Not having a familiarity with mushrooms that once thrived in an ecosystem that has since radically undergone desertification is understandable.

Unfortunately, scientists from other disciplines often have a difficult time recognizing what is obvious to mycologists. This has been a repeating pattern that has prevented the knowledge of psilocybin mushrooms from being part of the recognized historical record. I caution skeptics to be more circumspect in denying these forms could be mushrooms. One Egyptologist I consulted insisted that they could not be mushrooms but were upside down shovels, which is probably academically safer than to say they look like mushrooms. I think this underscores an inherent problem archaeologists and anthropologists face: that when the ecosystem changes and, in this case, desertification encroaches, once commonly encountered mushrooms in a verdant river delta are now nearly extinct in that arid bioregion. This lack of mycological knowledge naturally confounds and biases many modern experts' interpretations of ancient psilocybin mushroom practices.

It should not be surprising that when psilocybin mushrooms were preserved in honey or used with other herbs they would ferment and create intoxicating brews. The Reinheitsgebot (Bavarian Beer Purity Act) of 1516 purportedly banned mushrooms or any other potentially psychoactive ingredient or additive from being used to brew beer. This law was probably enacted for

The blue lotus (*Nymphaea nouchali* var. *caerulea*) closes as evening approaches and reopens with the dawning of a new day. The blue lotus was revered by the ancient Egyptians as a spiritual sacrament, representing rebirth.

The author (left) with Christian Rätsch, PhD, holding *Psilocybe azurescens*.

many reasons, but one of the most important was to prevent the addition of hallucinogenic or even toxic adulterants. The act dictated that only water, barley, hops, and natural fermentation (from yeasts) could be used by brewers in making and marketing beverages labeled beer. Limiting ingredients to barley and hops supposedly conserved grains like wheat and oats for making bread. The late Christian Rätsch, a German ethnopharmacologist, suggested this was one of the first European drug laws.[42] By coincidence or design, another pagan practice was likely suppressed, but the fog of history also makes the reasons for doing so difficult to ascertain.

Xōchipilli, the Prince of Flowers, a sixteenth-century Aztec statue from central Mexico, features several psychoactive plants and mushrooms. Note the *Psilocybe* mushrooms with inrolled margins at the base and on his kneecaps.

New World Psilocybin Mushroom Use: A Documented History

Psilocybin mushroom use in the New World has been well documented over several hundred years, based on both archaeological discoveries and written descriptions. One of the most profound discoveries, which was unearthed in the mid-1800s at the base of the Popocatépetl volcano in central Mexico, was a statue that has become known as Xōchipilli, the Prince of Flowers, the Aztec god of love, dance, agriculture, and visionary plants. Estimated to have been carved in the 1500s, the statue reflects the deep appreciation and respect the Aztecs afforded to visionary plants and psilocybin mushrooms. Currently displayed at the Museo Nacional de Antropología in Mexico City, the base and the kneecaps of Xōchipilli have phenotypically accurate carvings of what could be *Psilocybe caerulescens* or *Psilocybe aztecorum*. The Aztecs used psilocybin for divinatory advantages: seeing into the future, finding lost objects, diagnosing diseases, and voyaging into the spirit world.

Historians have also been fascinated by *teonanácatl*, a concoction that was served at the crowning of the Aztec emperor Ahuitzotl in 1486. Diego Durán, a Dominican friar, reported that inebriating mushrooms were eaten at the great festivity honoring the accession to the throne of Moctezuma II, the famed emperor of the Aztecs, in 1502. These reports were documented in the *Florentine Codex* by Spanish missionary priest Bernardino de Sahagùn, circa 1529. Not for several hundred years thereafter do we find another written record by Western observers, although the traditional use of psilocybin mushrooms undoubtedly continued. In his 1885 dictionary of the Nahuatl language, spoken in Oaxaca, the French linguist Rémi Siméon mentioned the use of hallucinogenic (presumably) psilocybin mushrooms: "Teonanacatl, espece de petit champingnon qui a mauvais gout, enivre et cause des hallucinations." [Teonanacatl, a kind of small mushroom that tastes bad, intoxicates and causes hallucinations.][43]

The name *teonanácatl* translates as the "flesh of the gods" (the same translation was attributed to mushrooms by ancient Egyptians). The fourteenth-century Mixtec document *Codex Vindobonensis Mexicanus I* includes repeated mushroom-shaped images, including one of seven gods clasping two or three mushrooms. The *Codex Vindobonensis* was presented as a gift to King Charles V of Spain in 1519, and it is currently housed at the Austrian National Library in Vienna.

In February of 1519, a flotilla containing the initial invasion force of 500-plus Spanish conquistadors arrived on the coast of the Yucatán Peninsula in Mexico, led by Hernán Cortés with eleven ships, a multitude of cannons, and more than a dozen horses as well as cattle, sheep, goats, and European plants. Accompanying this expeditionary force were Catholic priests to convert the Aztecs and numerous scribes to record this historical encounter. One of the missionary priests was Bernardino de Sahagùn, who documented the use of psilocybin mushrooms by the Aztecs. Publishing his observations in a 1529 document that has come to be known as *The Florentine Codex*, de Sahagùn recorded the use of intoxicating mushrooms: "These they ate before dawn with honey . . . when they began to get heated from them, they began to dance, and some sang, and some wept. . . . Some cared not to sing, but would sit down in their rooms,

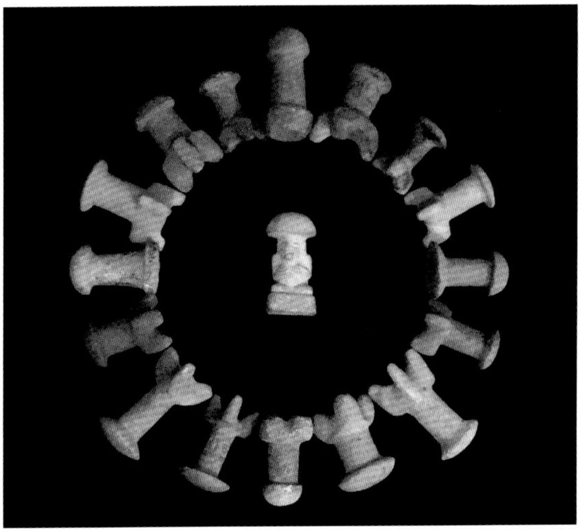

Mayan mushroom stones from 1500 BCE to 500 CE hint at the use of psilocybin mushrooms. A few have depressions at the center, suggesting that they also served as metates for grinding substances. Many mushroom stones were unearthed by Mayan villagers, who sold them at roadside stands to tourists visiting Guatemala in the 1960s and 1970s. Others came into the art market when museums, in the act of deaccessioning, sold them to fund operational costs. Difficult to authenticate, numerous fakes have been offered for sale. Much rarer mushroom stones carved from jade have also been found from the Olmec culture, a pre-Mayan and pre-Aztec culture best known for carving giant basalt heads of their chiefs. More than 600 of the many thousands of stones that must have existed have been found thus far.

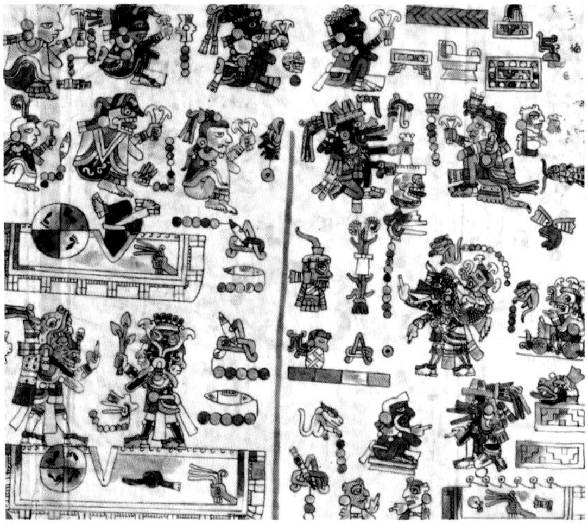

This *Codex Vindobonensis Mexicanus I* facsimile, the original dating possibly as far back as the fourteenth century, depicts multiple examples of Aztecs consuming psilocybin mushrooms. Of particular interest to me as a mycologist are the three conic-shaped, yellowish, *Psilocybe*-like mushrooms held by the figure in the lower left, bearing a strong morphological resemblance to *Psilocybe zapotecorum* or *Psilocybe yungensis*.

and stayed there pensive-like.... Then when the drunkenness of the mushrooms had passed, they spoke one with another about the visions that they had seen."[44]

During the Christian conquest of other cultures, Indigenous practitioners were commonly persecuted. Effigies representing their spirit world or gods were systemically destroyed as idolatry. In the decades following Cortés's 1519 arrival in Veracruz, Mexico, hundreds of thousands of Spaniards followed, seeking gold and destroying Indigenous culture and religion. The arrival of *Psilocybe cubensis* to the Americas likely happened via the cattle imported by Christopher Columbus to the Caribbean and/or later by the waves of Spanish conquistadors and agrarian-focused priests.[45] Although the Spanish likely brought a new psilocybin mushroom species to the region, ironically, they suppressed the use of psilocybin mushrooms generally. Psilocybin mushroom use was forced underground. History is written by the victorious, who also became censors. Subsequently, several hundred years followed with scant to no written records of psilocybin mushroom use, although the mushroom rituals endured.

The Modern Rediscovery of Psilocybin Magic Mushrooms

The rediscovery of psilocybin mushrooms by Western scientists and intellectuals had many twists and turns, illustrating how scientific opinions change over time. In the history of psilocybin use, many experts of this eclectic knowledge have kept their personal experiences secret so as not to risk jeopardizing their religious, political, economic, or academic standing. Some reports are noticeably circumspect.

For instance, in Charles McIlvaine's *One Thousand American Fungi* published in 1900, he reported witnessing several psilocybin mushroom journeys by ingesters of a dung-dwelling *Panaeolus*, which was identified (albeit likely in error) as *Panaeolus papilionaceus*. He stated, "The effects of *P. papilionaceus* are very uncertain. I have seen it produce hilarity *in a few instances* [my italics], and other mild symptoms of intoxication.... Being of small size, and not a prolific species, quantities of it are difficult to obtain."[46]

I wish I could go back in time 125 years and interview McIlvaine about witnessing a "few instances" of mild intoxication. I suspect he was underreporting. It stretches the imagination why anyone would seek this species for culinary purposes given that the mushrooms are, as he admits, small, and other robust species of culinary value would be more inviting, such as

Agaricus campestris, the meadow mushroom that commonly grows in the same ecosystem.

In 1923, Blas Pablo Reko, an Austrian medic and ethnobotanist living in Mexico, wrote to the U.S. National Herbarium that teonanácatl was not peyote as the American ethnobotanist William Stafford had claimed, but was, in fact, an intoxicating mushroom. In 1936, anthropologist Robert J. Weitlaner was given mushroom specimens by his Indigenous contacts who had collected them near the southern Mexican town of Huautla de Jiménez in the state of Oaxaca. Weitlaner sent these specimens to Reko, who then forwarded them to the Botanical Museum of Harvard University. In 1938, 23-year-old ethnobotanist Richard Evans Schultes traveled from Harvard to Mexico, where he joined forces with Reko to investigate further. They also visited Huautla de Jiménez, where they learned more about the use of the sacred mushrooms but were not invited to participate in any veladas—traditional mushroom ceremonies.[47]

Also in 1938, Jean Bassett Johnson and a group of anthropologists witnessed a secret nighttime mushroom ceremony in Huautla de Jiménez.[48] They had received information from Eunice V. Pike, an American missionary who belonged to the Wycliffe Bible Translators.[49] Johnson and his eventual wife, Irmgard Weitlaner-Johnson, daughter of Robert J. Weitlaner, reported that "the brujo (sorcerer) is under the influence of the narcotic mushroom, it is the mushroom which speaks, and not the brujo." From their extensive interviews, Eunice Pike and fellow-missionary Florence Cowan learned that the Mazatecs believed the mushrooms formed where the tears and blood of Jesus Christ fell to the earth.

"The mushrooms say that she is the child god, a spirit, a saint, a creator, the heart of Christ, the heart of the Virgin Mary. She is light, speech, thought." Pike and Cowan lamented, "What teaching must accompany the introduction of the Lord's Supper to the Mazatecs to prevent the Mazatecs among them from seeking the same experience from the bread and wine that they had formerly gotten from the mushroom—and being disappointed? Some Mazatec believers, chiefly new Christians, have continued to use the mushroom for a while. Arguments against it have accomplished very little." [50]

Synonymous with the body of Christ, they believed the mushrooms should never be cooked as that would be equivalent to harming God. The Mazatecs with whom Pike and Cowan spoke claimed they saw heaven when they ingested the mushrooms and seemed perplexed that the Christians could not also see heaven, but only had the promise that they would see heaven after death. The two missionaries admitted to not having a convincing response.

Syncretism between the religious practices of Indigenous Mazatecs and Catholicism—the obvious similarity between consuming mushrooms and the Eucharist—became entrenched in Mazatec traditions post European contact. This Indigenous-Catholic invocation represents the fusion of two cultures in the Mazatec mushroom veladas. Despite their efforts, Catholic priests and missionaries were frustrated in their attempts to remove psilocybin mushrooms from this newly woven cultural fabric of the Indigenous Mazatecs. Moreover, the Mazatecs believed in only planting crops where mushrooms were present because they knew that soil would be rich (and barren where they didn't find mushrooms). That mushrooms nourish the land is another example of the remarkable wisdom that these Indigenous people had. Psilocybin mushrooms remain transcendent—bridging cultures with divine respect for nature and the spiritual universe whose presence guided them to enhance agricultural outcomes.

Specimens in Schultes's *Panaeolus* collection were originally described as having a cuspidate cap and thus appear to have been subsequently misidentified; instead, they were likely *Psilocybe mexicana*. *Panaeolus papilionaceus* is typically psilocybin-inactive but it can co-occur with *Psilocybe mexicana*, since both are grassland species thriving where cows, horses, and sheep graze. These species can often be found in grassy areas where dung has decomposed, melting into the surrounding grasses and obscuring the obvious dung habitat wherein it dwells.

The dung-loving *Panaeolus papilionaceus* (= *Panaeolus campanulatus* and *Panaeolus sphinctrinus*) has confused many experts. Analyses have found some collections of *Panaeolus papilionaceus* to have low levels of psilocybin, but inconsistently. The records have been difficult to decipher. These mix-ups are common when the taxonomy of a species is being determined, and *Panaeolus* species have a history peppered with misidentifications, mixed collections, and lost voucher collections. Misidentifications by experts is part of the learning process in deciphering new species. Reexaminations using DNA analysis has helped to resolve some, but not all, of the confusing history associated with psilocybin mushroom classifications. Of concern is that many mushroom vouchers in herbaria are mixed collections of more than one species. Unless all the specimens are studied, experts can reach wrong conclusions, which other experts will promulgate. To this day, mushroom taxonomy is a fragile, nascent science.

I once bioassayed about thirty raw *Panaeolus papilionaceus* (they had a disgusting taste) in hopes they were psychoactive. They were not, but in the subtropics, the often co-occurring *Panaeolus cyanescens* and *Psilocybe cubensis* are psilocybin-active. More *Panaeolus* species, like the horse dung–loving *Panaeolus cinctulus*, once known as *Panaeolus subbalteatus*, can also co-occur with *Psilocybe cubensis*. (Both are commonly found in horse and cow pastures.) In fact, *Psilocybe mexicana*, *Panaeolus papilionaceus, Panaeolus cyanescens*, and *Panaeolus cinctulus* can all grow in the same fields and occasionally two species sprout from the same pile of dung. Each of the four mushrooms mentioned can be an indicator species for another; meaning that if you find one, you are likely in the right habitat for finding the others.

The Butterfly Panaeolus, *Panaeolus papilionaceus*, is psilocybin-inactive, often co-occurring with psilocybin-active mushroom species, presented here on cow dung. It is common to find many species of *Panaeolus* growing in proximity, which has led to much confusion about their taxonomy. This mushroom does not expand to plane, and the distinctive denticulate remnants along the cap edge are often missing due to weather or age. See also page 127.

In 1940, Richard Evan Schultes published an important paper in the *American Anthropologist* entitled "Teonanacatl: The Narcotic Mushroom of the Aztecs."[51,52] The specimens provided to him, which originated in pastures, were in poor condition and tentatively identified as belonging to the *Panaeolus campanulatus* var. *sphinctrinus* group. Once Schultes published his article, and for several years thereafter, Western scientists (including Schultes) carried forward this misidentification of teonanácatl (flesh of the gods) as *Panaeolus campanulatus* var. *sphinctrinus*, now known as *Panaeolus papilionaceus*. (Misidentification of mushrooms in science is not uncommon.)

However, I agree with other experts that teonanácatl was *not* a *Panaeolus* but rather a *Psilocybe* or a collection of *Psilocybe* species: most likely *Psilocybe mexicana, Psilocybe caerulescens*, and/or *Psilocybe aztecorum*. Subsequent collections of so-called magic mushrooms by the Mazatecs that were in better condition were identified by the French mycologist Roger Heim. The reexamined collections included not only *Psilocybe mexicana* but also the psilocybin-potent *Psilocybe caerulescens* and *Psilocybe cubensis*, then known as *Stropharia cubensis*.

These are the earliest reports in the Western scientific literature describing the rediscovery of psilocybin mushroom use in Mexico—mushrooms that were known and used by Indigenous mycologists for hundreds, if not thousands, of years beforehand. In the United States, mycologists such as Charles McIlvaine knew in the late 1800s that a species resembling *Panaeolus papilionaceous* caused intoxication, but there are no prior written records known to this author of European immigrants intentionally consuming psilocybin mushrooms in North America.

Nearly 20 years later, after Richard Evans Schlutes first learned of visionary mushrooms, a Mazatec medicine woman from Oaxaca named María Sabina Magdalena García guided self-described ethnomycologist R. Gordon Wasson and some of his entourage on a psilocybin mushroom journey. Wasson was the vice president of public relations at JP Morgan Bank and, along with his physician wife, Valentina (Pavlovna) Wasson, spent decades researching the use of hallucinogenic (primarily *Psilocybe* and *Amanita*) mushrooms across many cultures.

The Wassons' entry into the field of ethnomycology arose after the two had diametrically opposite reactions to a chance encounter with wild mushrooms on a belated honeymoon after their marriage in 1926. Valentina (Tina) grew up collecting and consuming mushrooms in her home country of Russia before departing during the Russian Revolution and becoming a physician in London. She knew many mushroom species by name. One day on their honeymoon in the Catskills Mountains of southeastern New York State, they chanced upon a group of majestic mushrooms. She was delighted with their find, while Gordon was aghast at the thought of eating them. They were fascinated by how each other's cultural upbringing contributed to their reactions, which they later termed *mycophilia* and *mycophobia*, and both were determined to learn more. This experience sparked a lifelong exploration of how different cultures view and use mushrooms, eventually leading them to the deep historical use of psilocybin mushrooms in the New World.

Gordon and Tina's interests eventually led them to Oaxaca through contacts provided by Richard Evans Schultes, Blas Pablo Reko, and Robert Weitlaner. In 1953, the Wassons, presenting themselves as anthropologists, met a local family in Huautla de Jiménez who gave them shelter for the night. The next day, their hosts collected the sacred mushrooms nearby and showed them to the visitors, who soon thereafter attended their first ceremony with the shaman Aurelio Carrera. These Westerners apparently only *observed* the mushroom velada ceremony.[53] Gordon Wasson subsequently self-experimented with a mixture of wild mushrooms but disappointingly felt no effects. Writing in their book, *Mushrooms, Russia, and History*, the Wassons described their experience: "As the days went by, we felt increasingly disappointed that we were not amassing an abundance of the sacred mushrooms. The early ending of the rains had made them scarce. RGW [R. Gordon Wasson] depleted our precious store by eating three of the small ones and one of the large, but more could not be spared. Bitter to the taste, they were not sufficient to cause psychic symptoms."[54]

Upon return visits, one of which was 2 years later, the Wasson party met Cayetano García Mendoza, who was employed by the local municipality as the Junior Mayor. He introduced them to his relative María Sabina, a Chojn Chijné, or wisdom keeper, for another ceremony. These encounters and sharing were initially mutually agreed upon, but secrecy was to be respected. It was not until Gordon's third visit to the Mazatecs in 1955 that he was invited to fully participate in a velada. On the night of June 29, 1955, Wasson and his colleague Allan Richardson participated in a sacred psilocybin mushroom velada with María Sabina with specimens collected by their guides from decomposing sugar cane bagasse nearby. María Sabina extolled the mushrooms' appearance with praise and song.

Wasson and his expanded entourage attended another velada with María Sabina on July 2, 1955. This time Wasson consumed six pairs of what appeared to be *Psilocybe caerulescens* over 30 minutes, clearly a hero's journey dose. Richardson did not partake this time so he could focus on taking pictures, stating that "once was enough."[55]

Tina Wasson overlooks Gordon Wasson the day after ingesting a high-dose of *Psilocybe caerulescens* and journeying with María Sabina in Oaxaca.

A cespitose grouping of young *Psilocybe caerulescens*, a potent psilocybin species important in Indigenous Mazatec rituals.

María Sabina with her students. María Sabina insisted she was not a curandera but a *sabia*, a knowledge keeper or sage.

María Sabina in front of her altar, with family and R. Gordon Wasson.

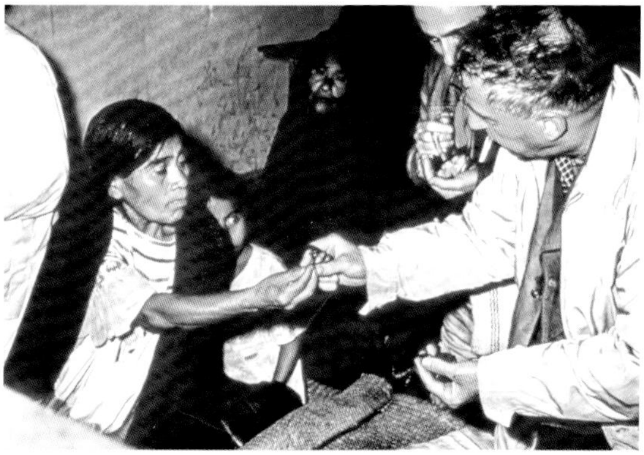

María Sabina gives R. Gordon Wasson pairs of *Psilocybe caerulescens*, a potent psilocybin mushroom.

María Sabina deep in ceremony.

Describing his experience that evening, Wasson said,

We were never more wide awake, and the visions came whether our eyes were opened or closed. They emerged from the center of the field of vision, opening as they came, now rushing, now slowly, at the pace our will chose. They were in vivid color, and always harmonious. They began with art motifs, angular such as might decorate carpets of textiles or wall paper of the drawing board of an architect.... Later, it was as though the walls of our house had dissolved and my spirit had flown forth, and I was suspended in mid-air viewing landscapes of mountains, with camel caravans advancing slowly across the slopes, the mountains rising tier above tier to the very heavens.... There I was, poised in space, a disembodied eye, invisible, incorporeal, seeing but not seen.[56]

A few days later, on July 5, 1955, and with a separate Indigenous guide, Tina and the Wassons' 19-year-old daughter, Masha, also ingested the sacred mushrooms. The phrase *magic mushrooms* was first coined by the editors of *Life* magazine on May 13, 1957, when they published Gordon's account of participating in a guided journey with a *curandera* (María Sabina), giving the Western world a glimpse of ancient psilocybin mushroom rituals.[57] (Allan Richardson's images of both sessions with María Sabina were published in the article.) The article alerted millions of Americans that magic mushrooms did exist and that the Indigenous peoples of southern Mexico knew how to use them for religious and therapeutic purposes. This public unveiling catalyzed the revolution of re-awareness that soon swept across the rest of the world.

A week later, on May 19, 1957, Tina published her story, "I Ate the Sacred Mushrooms," in nearly 12 million copies of *This Week*, a nationally

Unbeknownst to Gordon and Tina at the time, Gordon's return trip to Mexico in 1956 was partially funded by the CIA's Project MK-ULTRA, which was actively searching for new drugs for interrogations that could induce confessions.[58] A CIA document dated March 27, 1956, indicates that he (presumably Wasson) is "uncleared and unwitting of U.S. government interest in this project." (Wasson later reported in his diary that he was surreptitiously contacted by James Moore of the CIA.[59]) The document further notes that "he" had done three expeditions to date of a total of approximately ten planned field expeditions. The funding request was for $2,000 (which in today's dollars would be equivalent to $20,000 to $30,000). Comparatively speaking, this level of funding does not appear to have merited this expedition as a high-priority project. The money allocated was to reimburse photography and incidental expenses. It seems to me more like a fishing expedition for pursuing edgy ideas.

If we are to believe this partially redacted, declassified document, Wasson did not know that a CIA contractor was accompanying him on his expeditions and/or if he did later know, he kept this information private. I don't know of any further evidence that Wasson was aware at the time of his expeditions that the CIA had partially bankrolled his efforts. Later, he indeed learned that he had unwittingly become an asset for the CIA. And the Wasson expeditions were soon thereafter to create a public sensation that reverberates to this day.

syndicated Sunday newspaper supplement.[60] In it, she noted that psilocybin mushrooms might have therapeutic potential for Western medicine, an important and oft-overlooked milestone in psilocybin history.

The German-born mycologist Rolf Singer also traveled to Mexico to follow contacts provided by Richard Evans Schultes and Gordon Wasson and to collect specimens for further research. Singer met with Wasson in Huautla de Jiménez. In his book *Agaricales in Modern Taxonomy*, with volumes published in 1949 and 1951, Singer described the intoxicating mushrooms of Mexico not as a *Panaeolus*, but as a *Psilocybe*.[61] In 1958, Singer co-published a paper with Alexander Smith of the University of Michigan, naming several new bluing species of *Psilocybe* from the Pacific Northwest of North America, Mexico, Bolivia, and Argentina: *Psilocybe baeocystis*, *Psilocybe*

On May 13, 1957, *Life* magazine revealed to millions of readers the properties of psilocybin mushrooms, their cultural use by the Mazatecs, their features, and the Wasson team's deep journey with them.

R. Gordon Wasson and French mycologist Dr. Roger Heim hunting for psilocybin mushrooms in Oaxaca, 1955.

strictipes, Psilocybe collyboides, Psilocybe candidipes, Psilocybe aggericola, Psilocybe muliercula, and *Psilocybe yun-gensis.*[62] Singer had taken the unusual step of paying *Mycologia* (using money provided by Sam Stein, MD, director of the Chicago-based Bertram and Roberta Stein Neuropsychiatric Research Program Foundation[63]) so his article could be published out of order of receipt, to get ahead of Roger Heim's and Roger Callieux's planned publication of *Psilocybe wassonii.*

Heim published the name *Psilocybe wassonii* later in 1958 in a short paper entitled "Diagnose latine du *Psilocybe wassonii* Heim, Espece Hallucinogene des Azteques" [Latin description of *Psilocybe wassonii* Heim, a hallucinogenic species from the Aztecs],[64] but this name was subordinated by Singer and Smith's prior publication. Singer and Smith's publications also pre-empted some of the other would-be names that Heim and Callieux brought forward in their 1960 publication "Nouvelle contribution á la connaissance des Psilocybes hallucinogénes du Mexique" [Novel contributions on the knowledge of the hallucinogenic Psilocybes of Mexico].[65,66,67,68] (In taxonomy, whoever publishes new Latin binomials first gets the naming priority.) Clearly these four mycologists—Singer, Smith, Heim, and Callieux—were highly competitive. This conflict of the naming of *Psilocybe wassonii* and these other species created a rift, which is not uncommon in academia between brilliant minds eager to be first. I raised this issue with Alexander Smith, who told me he was unaware of this controversy until after publication. He said that he simply contributed microscopic and taxonomic comments upon receiving specimens from Singer. Singer, Smith, Heim, and Callieux were some of the most prolific mycologists in history, naming hundreds of mushroom species.

Tina Wasson died on December 31, 1958, at the age of fifty-seven. In every lecture by Gordon Wasson that I attended, he gave abundant credit to Tina, acknowledging her pivotal role in inspiring him on his subsequent decades-long exploration of ethnomycology. He further honored her memory when he later donated his books and artifacts to Harvard and named his gift The Tina and Gordon Wasson Ethnomycological

Psilocybe caerulescens being held by an ecstatic Xōchipilli, the Prince of Flowers. This reproduction is currently being used ceremonially by a Mazatec knowledge keeper.

Two-Eyed Seeing: Building a Cultural Bridge of Cooperation and Respecting Indigenous Practices

Any discussion of psilocybin mushroom use must acknowledge Indigenous rights and practices. The persecution, suffering, and exploitation of Indigenous peoples across the world is a painful and difficult history we must all recognize. Although we cannot change the past, we can change the future. Our diversity can be our greatest strength. The voices of *all* Indigenous people must be heard and respected. Now, with DNA testing, many of us are discovering that our genetic history is much more diverse—and complicated—than we knew.

Two-eyed seeing is an Indigenous-suggested model for us as a community to consider. This concept was first presented by Mi'kmaw Elders Albert and Murdena Marshall from eastern Canada to synergize Indigenous and non-Indigenous points of view.

Two-eyed seeing refers to learning to see from one eye with the strengths of Indigenous ways of knowing and from the other eye with the strengths of Western ways of knowing and using both eyes together.[69] The concept is simple: One eye sees from the perspective of Indigenous wisdom, practices, and worldview; the other sees from a Western perspective using modern technologies that can help to preserve, protect, and expand Indigenous cultural use. With two eyes, our collective vision is better than one. This method helps bridge cultural divides.

The National Science Foundation recently awarded a $30 million grant for establishing the Center for Braiding Indigenous Knowledges and Science (CBIKS), coordinating eight research hubs in the United States, Canada, Australia, and New Zealand based on the philosophy of two-eyed seeing. This is the path that many are embracing as we further explore and expand the knowledge of using psilocybin mushrooms for spiritual and medicinal purposes. The message from the spirit of psilocybin mushrooms is to share. Our planet is suffering. We have a crisis in creativity. We all lose if we do not collaborate. We must protect Indigenous practices as medical science potentially builds upon Indigenous healing practices. As Snuneymuxw First Nation Elder Geraldine Manson from western Canada stated, now is the time for *Naut sa mawt* ("working together, as one mind and spirit"). We cannot change the past, but we can learn from it to forge a better, collective future.

Collection. The Mycological Society of America created an award in the Wassons' honor. In 2015, I became the first recipient of The Gordon and Tina Wasson Award in recognition of my contributions to the field of mycology, which is one of my greatest accolades.[70]

Although Wasson disguised the name of María Sabina and her village in the *Life* article, he revealed her true identity and location the same year in the limited print run of 512 copies of volume 2 of his and Tina's book, *Mushrooms, Russia, and History*.[71] That Wasson would break confidentiality with María Sabina is seemingly in contradiction to the reverence and respect he espoused for her and the Mazatecs. Once people learned María Sabina's name, thousands flocked to Huautla de Jiménez to seek her out—sadly, to her detriment and that of her village. The spiritual rituals that had been hidden from Westerners since the persecutions of the Spanish conquest were exposed, and the rush of people, money, and notoriety proved to be disruptive. Oaxaca has not been the same since.

Interestingly, although María Sabina was a Mazatec wisdom keeper, she was also a devout Catholic. Her mushroom rituals were held with motifs of the Holy Trinity present. This fusion of Indigenous and foreign religious practice, known by anthropologists as syncretism, was perhaps a way for Indigenous peoples to maintain their traditional practices and survive oppression by the Spanish. There are traumatic histories throughout the world of Indigenous peoples whose use of psychoactive mushrooms and plants have been repressed. Today, the Mazatecs continue their sacred rituals, which are being reinvigorated by younger Indigenous generations continuing to practice and, in some cases, rediscovering long-held traditions. Because the demand for psilocybin mushrooms makes finding wild ones more challenging, the use of in vitro–cultivated *Psilocybe cubensis* is helping to provide the much-needed psilocybin mushrooms.

Once imprisoned for practicing psilocybin mushroom medicine, María Sabina is now heralded as a national hero in Mexico. Both María Sabina and Tina Wasson were dedicated to finding, identifying, and using mushrooms. They were not only medicine practitioners, but also mycologists. These two powerful women sparked the international psilocybin revolution. I will forever credit them for their example; they blazed the trail that we now follow. Most significantly, we owe a debt of gratitude and deep respect to the Mazatecs for keeping this ancient tradition alive in the face of sometimes lethal oppression by Spanish colonialists and the church.

The unveiling of the Mazatec traditions to the Western world is unfavorably viewed by many as an example of colonialism and extraction.[72] As psilocybin mushroom use escaped the confines of the Mazatec culture, it stimulated the revolution in interest we see today. But were the Mazatecs the only Indigenous peoples who used psilocybin mushrooms? Logically, many cultures throughout human history likely used some of the more than 220 psilocybin mushroom species known across the world today. The history of psilocybin mushroom use has been obscured by the persecution of Indigenous practitioners by invaders whose superior military technology, subsequent religious domination, and diseases drove practitioners of psilocybin ceremonies underground. That any records have survived is remarkable. (For further reading on the widespread historical examples of possible psilocybin mushroom use by Indigenous peoples, I recommend the annotated bibliography of Spiers et al.[73])

The first time I saw *Psilocybe* mushrooms being grown were in photographs of flasks from Roger Heim's Paris lab and in the 1958 Singer article showing gorgeous specimens of *Psilocybe cubensis* fruiting from compost. In 1960, mycologist Leon Kneebone from the Pennsylvania Agricultural Experiment Station, a center for supporting the button (*Agaricus*) mushroom industry, also published methods for cultivating psilocybin mushrooms—on agar, rye grain, sugar cane bagasse, and compost.[74] He received several specimens (*Psilocybe aztecorum, Psilocybe cubensis*, and *Psilocybe mexicana*) from Rolf Singer, who had collected the psilocybin mushrooms in Mexico. Kneebone's rarely referenced article became a tipping point for the widescale cultivation of *Psilocybe cubensis*. The sterilized grain method for fruiting *Psilocybe cubensis* was much simpler and easier than labor-intensive manure compost cultivation for home cultivators.

Notably, *Psilocybe cubensis* will spontaneously fruit on sterilized grains (i.e., aged spawn) whereas *Agaricus bisporus* (button) mushrooms will not. Growing mycelium on sterilized grain had been a common practice for decades as the preferred method of generating spawn for the inoculation of manure-based composts for the commercial production of button mushrooms.

Once Roger Heim, Alexander Smith, Rolf Singer, and Leon Kneebone had published their cultivation techniques, the world at large was exposed to a sustainable method for growing psilocybin mushrooms. This discovery revolutionized home cultivation of psilocybin mushrooms, including *Psilocybe* and *Panaeolus* species. Traveling to distant lands to find psilocybin was no longer the only option. People could stay at home and grow them. This newfound ability reduced pressure on wild harvesting in Mesoamerica and lessened the need for magic mushroom tourism.

With larger, more dependable supplies, psilocybin mushroom use steadily increased during the 1970s and 1980s. Currently, psilocybin mushroom consumption continues to trend upward, from baby boomers to zoomers (Gen Z), as scientific benefits are being validated. About 12 percent of respondents in a recent U.S. survey reported using psilocybin during their lifetime, with 3.1 percent having used psilocybin mushrooms in the past year. In this national survey, an estimated 8 million American adults used psilocybin in 2023, far more than any other psychedelic.[75] We do not yet have worldwide surveys, but I estimate the number of users to be surging to hundreds of millions.

Psilocybin Mushrooms in Their Natural Habitats addresses the important and growing societal need for accurate identification and safe dosing guard rails, to prevent harm and maximize benefits. The psilocybin mushroom movement is sweeping the world, awakening consciousness at a time that's critical for saving our species and ecosystems. When will we cross the threshold toward a mass paradigm shift in consciousness? For me and many others, this journey has deepened our spirituality and confirmed our shared unanimity of being. Psilocybin mushrooms could help elevate us to the next level as a species.

Psilocybin Mushroom Reports from the Pacific Northwest of North America

I live in the Pacific Northwest and spend most of my time hunting mushrooms from Northern California to British Columbia. The fungi of this bioregion are my focus and specialty, though I have traveled across the planet studying *Psilocybe* species. People *may* have been using psilocybin mushrooms in the Pacific Northwest for spiritual purposes in the distant past, but lifelong investigations by ethnobotanists like Nancy Turner and others have yet to document their use.[76] Native psilocybin mushrooms are extraordinarily rare and unobvious compared to the Fly Agaric (*Amanita muscaria*), a large, bright red mushroom that is glaringly obvious and has been used by Indigenous peoples worldwide, including in the Pacific Northwest. Psilocybin mushrooms can be especially hard to find in the rich mycodiversity of this bioregion, as they are hidden among thousands of other species growing in the same habitats. However, their rarity does not mean they were never used. The loss of an elder mycologist breaks the chain of knowledge. These mushrooms may have been discovered, lost, and rediscovered many times over generations. Although written records of psilocybin use by Indigenous residents have not yet come to

The caps of the Liberty Cap, *Psilocybe semilanceata*, undergo a radical, progressive color change. At first the moist mushrooms are dark chestnut brown, but lighten in color to a pale yellow brown as they dry. This state of progression of cap discoloration is called hygrophanous.

my attention, I was once told by an Indigenous Nisqually friend that their grandparents spoke of using small psilocybin-like mushrooms ceremonially for invoking visions.

Although psilocybin mushrooms were being collected in Oaxaca by foreigners soon after the Wassons' *Life* magazine story was published, the psilocybin species of the Pacific Northwest escaped widespread attention until two decades later. The earliest written records in English that I have found of psilocybin mushroom ingestion in the Pacific Northwest were accidental "poisonings" of two families near Milwaukie, Oregon, and Kelso, Washington, in 1960 and 1961.[77] The mushrooms were supposedly *Psilocybe baeocystis*, but the black-and-white photograph used in one of the publications is clearly *Psilocybe cyanescens*.

There is another recorded incident of psilocybin mushroom use from the 1960s, but this time the ingestion appears to have been intentional. Royal Canadian Mounted Police arrested a student at the University of British Columbia for possession of Liberty Cap (*Psilocybe semilanceata*) in 1965. And in the 1970s, the Masset grassy region of the Haida Gwai'i islands in British Columbia became a mecca for seekers of so-called magic mushrooms. Hundreds of hippies traveled to this remote island to hunt for Liberty Caps. In a 2012 article published in the Vancouver Mycological Society, the authors speculated that the Liberty Caps were brought over from Europe with grazing livestock and the fodder plants needed to support them.[78]

In the 1960s and 1970s, there was much confusion about *Psilocybe baeocystis* and *Psilocybe cyanescens*. In a 1977 article, Dr. Andrew Weil reported on the rapidly emerging popularity of picking psilocybin mushrooms in Oregon, especially those growing on wood chips.[79] Although he was told by local mycologists that the species popularly collected was *Psilocybe baeocystis*, it was likely *Psilocybe cyanescens*. Back then the taxonomy of *Psilocybe* was unclear because few experts could identify them accurately. *Psilocybe baeocystis* is a rare species even today, while *Psilocybe cyanescens* remains common. The insights, experiences, and courage of this Harvard-trained physician greatly influenced me and others who were exploring psilocybin. Dr. Weil's significant contributions to the emerging psychedelic renaissance continue to this day. Like María Sabina and Tina Wasson, Andrew Weil is part of the continuum of revered knowledge keepers who expand modern medicine with his perspectives on psilocybin mushroom practices.

Along with other mycologists, I began to organize a series of mushroom conferences in

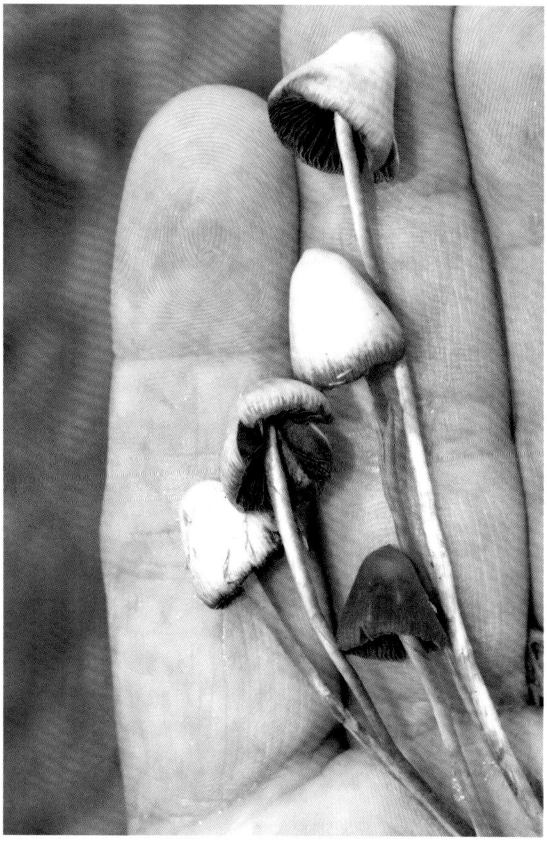

The Liberty Cap, *Psilocybe semilanceata*, is a distinctive grassland species, with a strongly hygrophanous cap, purple brown spores, and usually (but not always) adorned with a sharp nipple. This species occasionally has bluish tones, especially with the mycelium at the bases of the stems. It is very high in psilocybin, a stable molecule, and low in psilocin, which is unstable. Like *Psilocybe pelliculosa*, this species has a separable gelatinous pellicle (see page 181).

The Wavy Cap, *Psilocybe cyanescens*, is typically chestnut brown to caramel in color when moist. Here two mushrooms turn yellow as they dry out. Mycologists term this color transition hygrophanous.

the late 1970s in response to the interest in psilocybin mushrooms and because conservatively minded mycological conferences did not want to speak about them. At the age of twenty-one, my work was covered under a Drug Enforcement Administration license held by my professor, Dr. Michael Beug of The Evergreen State College, so we felt emboldened to organize gatherings of experts on this subject.

The first Hallucinogenic Mushroom Conference was held at Millersylvania State Park near Olympia, Washington, in 1976 and was attended by R. Gordon Wasson, Jonathan Ott, David Repke, Dale Leslie, Michael Beug, me, and many others. In 1979, Jim Jacobs, Jeff Chilton, Gary Menser, and I began the MycoMedia conferences that continued for another decade. The MycoMedia conferences focused on building bridges between academic mycologists, who were often not as well informed on psilocybin mushrooms, and experienced, non-academic ("amateur") mycologists. These conferences proliferated. The Telluride Mushroom Festival founded by Dr. Emanuel Salzman, Joanne Salzman, Dr. Andrew Weil, Gary and Irene Lincoff, Art Goodtimes, Lee and Linnea Gillman, myself, and others debuted in 1981. Psilocybin-focused conferences continue to thrive to this day. I see these gatherings as a continuum of the transfer of sacred knowledge that goes back thousands of years, across continents and cultures.

The Millennium Mushroom Conference in Breitenbush, Oregon, on October 31, 1999, was a meeting of the Merry Pranksters (Ken Kesey and friends with the bus, Further) and psychedelic scientists. Along with me, Ken Kesey, George Walker, Ken Babbs, Mountain Girl (Carolyn Garcia), Alexander and Ann Shulgin, Dr. Andrew Weil, Dr. Michael Beug, Kit Scates, David Aurora, Gary Lincoff, Christian Rätsch, Jonathan Ott, David Tatelman, Satit Thaithatgoon, Chris Kilham, Dr. Manny Salzman, Joanne Salzman, Tom Riedlinger, Beverly Jenden-Riedlinger, Dorothy Beebee, Miriam Rice, Steve Rooke, Manny, Joanne and Naomi Salzman, many other notable psychonauts attended.

Paul Stamets, Sasha Shulgin, and Terence McKenna in Palenque, Mexico, 1998.

MycoMedia's Mushrooms I conference was held in Siltcoos Station, Oregon, in 1979. Featured here are (left to right) Dr. Steven Pollock, James Jacobs, myself, Dale Leslie, Dr. Gastón Guzmán, Jeremy Bigwood, and Jonathan Ott.

Alexander Shulgin, Jonathan Ott, Christian Rätsch, and Andrew Weil were some of the faculty at the 1999 Millennium Mushroom Conference.

In 2023, in Denver, Colorado, Rick Doblin and the MultiDisciplinary Association for Psychedelic Studies (MAPS) organized the largest gathering of psychonauts known thus far: approximately 12,000 scientists, knowledge keepers, and the psychedelically curious. As it has been for thousands of years, the tradition of sharing knowledge among large groups continues. Psilocybin research was a principal focus.

Paul Stamets, LaDena Stamets, and Terence McKenna at the MycoMedia Conference in Breitenbush Hot Springs, Detroit, Oregon, in 1983. McKenna published several books and was a psychedelic leader in Northern California during the 1980s and 1990s.

Roland Griffiths, psychopharmacologist from Johns Hopkins University School of Medicine, published clinical studies using psilocybin, re-invigorating the scientific community to renew research on psilocybin for its benefits. He and his colleagues' 2006 and 2021 clinical studies are credited by many for legitimizing psilocybin as therapies.[80]

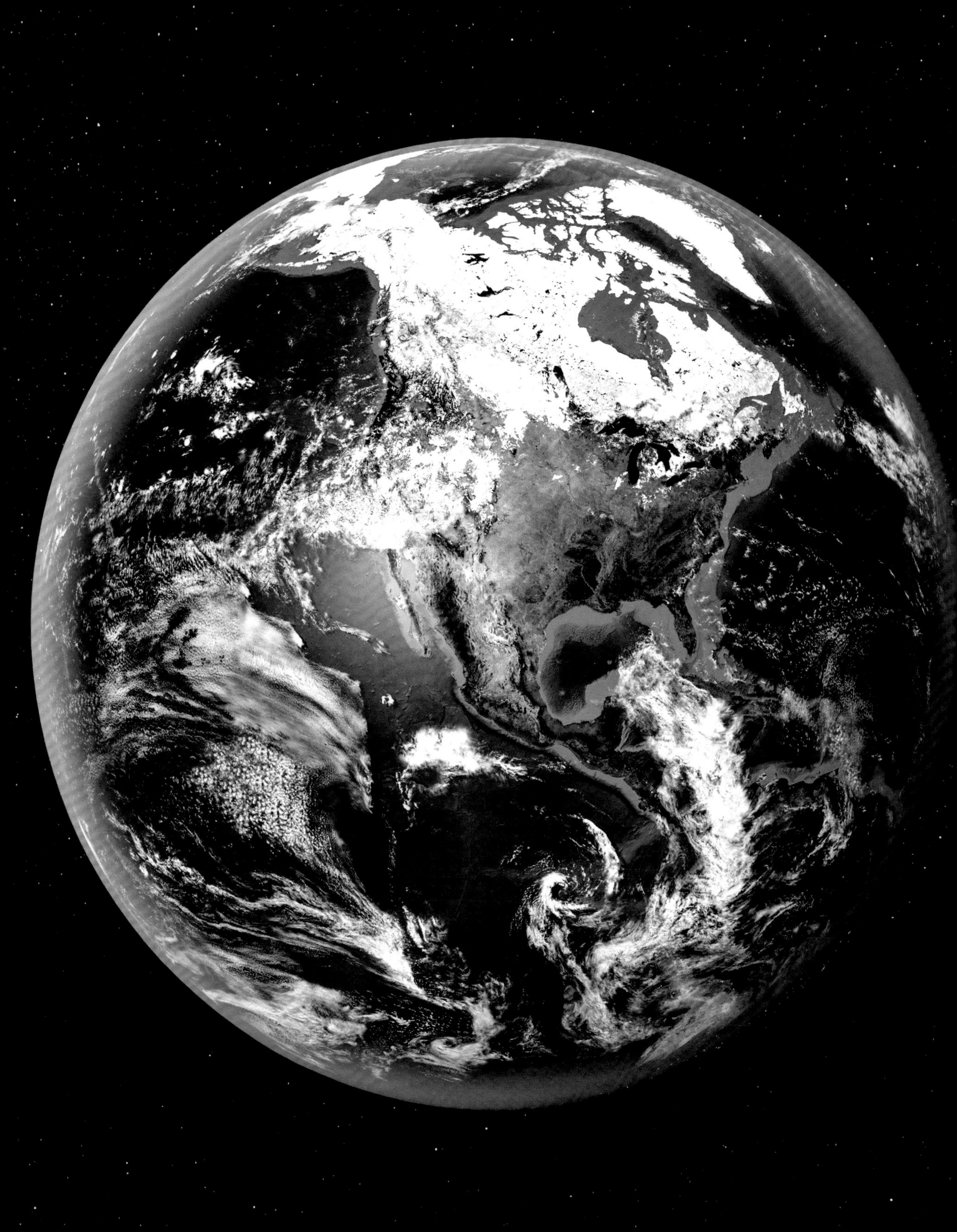

Chapter 2

Where to Find Psilocybin Mushrooms

Psilocybin mushrooms grow on every continent on Earth except—yet to be documented—Antarctica. More than 220 species of psilocybin-containing mushrooms have been described since the early 1800s. Most are widely distributed in two habitats: forests and fields. The vast majority of psilocybin mushrooms are saprobes (species that live on dead and/or decaying matter). The depth and diversity of psilocybin mushroom species in the library of nature offers a wide selection of candidates to engage sustainably since many can be cultivated.

Wood-Loving Psilocybin Mushrooms

In my bioregion, the only psilocybin-active species I have found in the old-growth Pacific Northwest fir forests is *Psilocybe pelliculosa*. Most wood-decomposing psilocybin mushrooms are found wherever forest debris is generated, especially in transitioning forest stands, where the land is disturbed. Many psilocybin species are found in riparian zones that undergo frequent flooding and carry trees, branches, and wood detritus downstream. Whether in the Pacific Northwest of North America or in the highlands of Mexico or Spain, these zones are often ideal places to look for *Psilocybe* species. In Oaxaca, Mexico, landslides, road cuts, trails, and agroforestry debris often can nurture the growth of *Psilocybe zapotecorum* and *Psilocybe caerulescens*. Curiously, comparatively few lignicolous (wood-dwelling) psilocybin mushrooms are found in pristine, undisturbed habitats. Most thrive where there is disturbance in the forests. Since humans are more skilled at generating debris fields than any other animal, it's no surprise that mushrooms appear where we disturb nature, especially where trees are cut, wood chips are spread for landscaping, or roads and trails are built through forests.

Another landscape where mushrooms thrive is flooding rivers. One excellent example is the Columbia River that flows between Oregon and Washington in the United States. The Columbia basin has a wide swath of flood-prone habitats for *Psilocybe cyanescens* and *Psilocybe azurescens*. The recurring influx of wood debris and

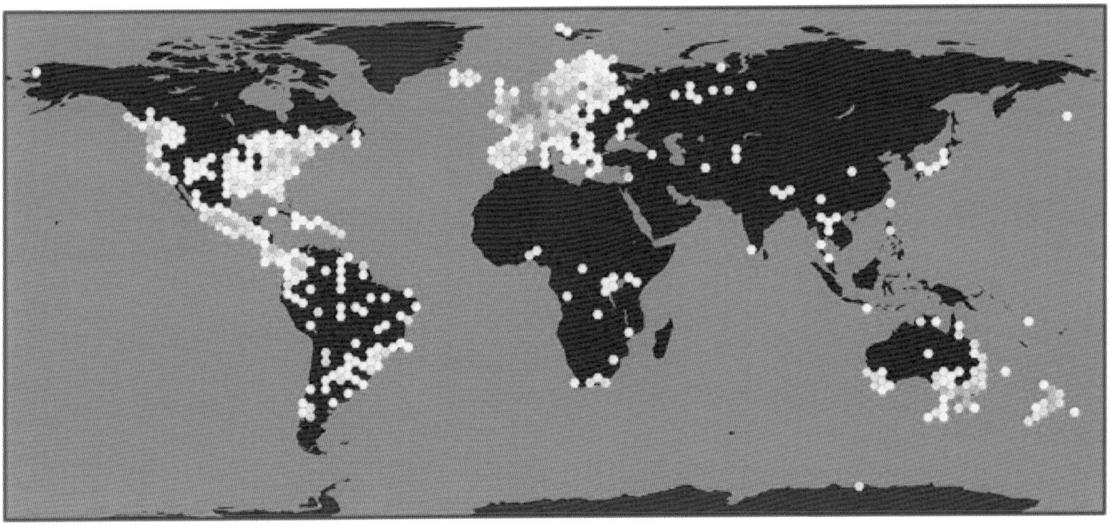

Map showing the distribution of more than 5682 psilocybin mushroom collections from 1800 to 2022, representing many of the approximately 220 known psilocybin-active species.

fluctuating water levels that exit an average of 265,000 cubic feet (7500 cubic meters) per second create shorelines that are ideal habitats for several psilocybin mushroom species to flourish, especially *Psilocybe azurescens, Psilocybe cyanescens, Psilocybe baeocystis*, and *Psilocybe stuntzii.*

Psilocybe azurescens and *Psilocybe cyanescens* occur in the same habitats and are closely related, although the natural range of *Psilocybe cyanescens* is more extensive. *Psilocybe azurescens* has a comparatively evenly shaped outer cap margin, often with an umbo (a raised area) in the center of the cap. *Psilocybe cyanescens* tends to have a wavy cap margin, but infrequently has an umbo. Also, *Psilocybe azurescens* can become much larger than *Psilocybe cyanescens.* Both are potent psilocybin mushrooms that stain bluish when bruised.

When the first Europeans sailed to the Americas, their wooden ships were colonized with mycelium. Once the wood rotted and was no longer seaworthy, the ships were often dismantled and salvaged for building materials. (See pages 215 and 216 for a photo of *Gymnopilus luteofolius.*) Carpenters would avoid the rotted boards, tossing them into a debris pile. But some would appear to be useful as lumber. Eventually, the mycelially impregnated boards would inoculate the landscape with species often new to that coastal environment.

Although *Psilocybe cyanescens* was first described by Elsie Wakefield from the Royal Botanic Gardens at Kew in 1946,[1] there is a strong likelihood that this species is not endemic to the British Isles and Europe. The Kew Gardens imported hundreds of species of plants, ferns, and mosses from across the world. Among the collections were many trees and shrubs from the Pacific Northwest, including Douglas fir, alders, wild roses, azaleas, and rhododendrons. Could the gathering of plants from the Pacific Northwest have transported *Psilocybe cyanescens* to Britain and Europe? This is a plausible theory and likely would not have been a one-time event but repeated many times with arriving living plant collections from afar.

Psilocybe cyanescens, the Wavy Cap, was first described in 1946 from the Royal Botanic Gardens at Kew in Richmond, England. But did *Psilocybe cyanescens* colonize England and Europe from the United States? Many mycologists think so.

Psilocybe azurescens, Azzies, naturally inhabits the coastal regions near the mouth of the Columbia River as it flows into the Pacific Ocean.

A colony of *Psilocybe cyanofibrillosa* growing underneath a rhododendron at the Rhododendron Species Foundation in Tacoma, Washington.

This kind of migration and colonization is called *panspermia*—the spreading of a species' genome to new habitats. It makes sense that psilocybin species hitchhike with migrating humans and their domesticated animals. However, finding the ancestral habitats of psilocybin and many saprobic mushrooms is a continuing challenge to evolutionary mycologists because territorial boundaries are no longer confined by transoceanic divides.

As humans migrated and colonized new habitats, we chopped down trees for fires and buildings and, more recently, used wood chips for landscaping. We have created ideal opportunities for these often-rare species to be discovered near our more populated areas. In nature, other animals, such as woodpeckers, squirrels, beavers, and other rodents, also generate wood chips. Beavers gnaw on and fell trees to construct their dams and lodges, which are replete with abundant water and wood shards and perfect environments for psilocybin mushrooms.

Clusters of *Psilocybe cyanescens* forming on naturalized wood chips used for landscaping.

I wonder if beavers are ecological vectors that enable certain wood-dwelling psilocybin mushrooms to enter domesticated zones? Similarly, woodpeckers are known to inoculate trees with fungi as they drill into the wood.[2,3]

I suspect that many wood-decaying psilocybin species provide an endophytic-like defense within the ecosystem of developing—and aging—trees. Many fungal decomposers are also endophytes (they make their home between living plant cells), which is part of the natural mycobiome that protects the community from quick-to-kill fungi and parasites. When a host tree is significantly changed by woodpeckers, beavers, humans, and other animals, it may signal to the mycelium within that it is time to produce a mushroom. Rarely, if ever, does any fungal species live in isolation in its ecosystem. Evolution results in guilds of collaborating organisms whose skill sets are complementary for their collective survival.

Squirrels not only collect tree seeds and munch on cones, but they also dwell in cavities, taking advantage of the hollows originally created by woodpeckers. Also, squirrels have notoriously poor memories about where they stashed seeds. What they cache and never eat is an opportune habitat for wood saprophytes. Cougars, bears, racoons, rodents, deer, moose, elk—all scratch trees, and these entrance wounds let saprophytic fungi into trees.

Since ants have a long coevolutionary history as mycelium cultivators, I would not be surprised if there were other undiscovered relationships between ants and fungi. Some flying insects lay eggs in maturing mushrooms as homes for their larvae. When hatching, newborn flies carry with them a collection of spores—not unlike pollination services provided by bees.

Most fungi use wind as a dispersal vector. With most mushroom species, two compatible spores must meet, mate, and then form fertile (dikaryotic) mycelium. Successful mating decreases as individual spores spread further away from the originating mushroom. In contrast, animals pick up dense spore payloads upon contact, so compatible spores in proximity to one another are more likely to mate successfully as compared to the dispersing effects of

I found this *Psilocybe baeocystis* growing directly from a Douglas fir seed cone. *Psilocybe baeocystis* is a potent but rare psilocybin mushroom of the Pacific Northwest of North America. When squirrels gather these cones, and then remove and stash their seeds, they might be spreading this and other species.

air spreading spores apart over distances. In essence, mushrooms have figured out how to benefit from animals as diverse as beavers to birds in spreading spores essential for their survival. Mushrooms have evolved many ways to spread their progeny. They follow us as we travel, and they use us and other animals to propagate.

In the Pacific Northwest, I have never found *Psilocybe cyanescens* in an intact, undisturbed alder forest. And yet, when alder trees are cut and chipped, *Psilocybe cyanescens* often grows from the wood chips the next fall. Is the mycelium resident within the living trees, awaiting the opportunity to form mushrooms as soon as this host is felled? How long is the mycelia enmeshed in these forests, and what roles do they play to help keep the ecosystems healthy? Is *Psilocybe cyanescens* an endophyte, protecting the tree and coexisting as part of the alder's mycobiomic immunity? When these alder forests are dramatically damaged, two psilocybin mushroom species are triggered into formation: *Psilocybe cyanescens* and *Gymnopilus luteofolius*. Likewise, when oak trees are chipped, many mushroom species suddenly appear, including *Psilocybe allenii*. Are these types of saprophytic mushrooms nature's way of helping damaged environments transition and begin healing? It's a common theme in mycology that ecological disturbance of forests stimulates mushroom formation. And most psilocybin mushrooms seem to favor newly disturbed habitats, edge runners at the interfaces of transitioning ecosystems.

Where is the original home of psilocybin mushrooms that are now widespread, like the *Psilocybe cubensis* group? Was it Africa or India? We do not yet know. My namesake, *Psilocybe stametsii*, found during an attempt to chart the mycodiversity of a protected Ecuadorian cloud forest,[4] is from one specific region. Unlike *Psilocybe cubensis*, I fear that this species could be wiped out from harvesting since only two

Deep in the Los Cedros Biological Reserve in Ecuador, Giuliana Furci of the Fungi Foundation, Cosmo Sheldrake, Robert Macfarlane, Cèsar Rodrìguez-Garavito, and Josè Cuevas walk with the judges from the Constitutional Court of Ecuador who protected this forest under the Rights of Nature article of the Ecuadorean constitution.

Giuliana Furci on the same trail that day as she collected the holotype of a new species, *Psilocybe stametsii*.[5]

specimens have been found to date. Or is it more plentiful than we currently know? The survival of *Psilocybe stametsii* would clearly be ensured if we could get it into culture and grow it. But out of an abundance of caution, we should be careful about harvesting these delicate and rare species. If you are seeking a psilocybin mushroom to use, consider one like *Psilocybe cubensis* or others that can be easily cultivated. Given the thousands of strains in nature and hundreds now in culture, cultivating *Psilocybe cubensis* is more sustainable.

In chapter 4, I'll explain step by step how to create a psilocybin mushroom patch in your backyard using *Psilocybe cyanescens* as the exemplary species. This method works for most of the lignicolous *Psilocybe* species. Once resident, this mushroom's short-lived fruitbodies can be cultivated for decades simply by replenishing the debris field annually.

Pastoral & Grassland Psilocybin Mushrooms

The other habitat where many psilocybin species grow is in grasslands: fields, pastures, lawns, and other grassy areas. The predominant species in cooler temperate regions is *Psilocybe semilanceata*, the famed Liberty Cap. The predominant species in tropical lowlands is *Psilocybe cubensis*, the Golden Top. Both habitats have several other psilocybin species that commonly co-occur. Finding a location of one species frequently means you have found a location of others.

The morphologies of most psilocybin mushrooms such as Liberty Cap and Golden Top are influenced by multiple factors, resulting in everything from minute to majestic fruitbodies. The genomic phenotype, substrate nutrition, height of grass, and weather are major influencers on the mushroom fruitbody. For instance, *Psilocybe semilanceata* can have short stems when growing in short grass or long stems when growing in tall grass, as seen on pages 183–184. The caps also tend to be larger when found in dense clumps of sedges. The same is true with *Psilocybe cubensis* and other species. Clearly, the length of the stem is highly variable and not a dependable taxonomic feature. Stems elongate so the caps can release airborne spores from their gills.

The easiest location to find psilocybin mushrooms in subtropical and tropical regions around the globe is manured grasslands. *Psilocybe cubensis*, a coprophilous (grows on animal dung) species, stands out; you can often see the golden tops from hundreds of feet away. I have enjoyed collecting them in Palenque, Mexico, in cow pastures surrounded by the jungle. In Texas and Florida, I could see their golden caps glistening in the sun from afar. In cow fields, these mushrooms are more abundant when nestled among tufts of grasses that provide shade and nutrients. *Psilocybe cubensis* is a lowland-loving species and less common above several thousand feet in elevation. In the highlands of Mexico, for instance, *Psilocybe zapotecorum* and *Psilocybe caerulescens* are widely collected and deeply rooted in ancient Mazatec tradition. Species have moved great distances thanks to animal and human migrations. Psilocybin mushrooms can be found during the rainy season in different regions of the world. Most are within generally defined temperature ranges called isotherms. Rainfall and air temperatures determine the rapid appearance of various mushroom species depending on the ecosystems in which they have evolved. Of course, species can be found outside of these ranges, but the following two representative isotherms are those where most of the psilocybin species dwell. I've listed eighteen representative species following, and some cross over these two isotherms. Day to night fluctuations and evaporative cooling from rains, wind, shade, slopes, and proximity to aquifers all influence the isotherms in which mushrooms thrive.

Pastoral, lowland habitat for picking Liberty Cap, *Psilocybe semilanceata*.

40 to 60°F (4.4 to 15.6°C): *Conocybula cyanopus* (formerly *Pholiotina cyanopus*), *Psilocybe allenii, Psilocybe azurescens, Psilocybe baeocystis, Psilocybe cyanescens, Psilocybe fimetaria, Psilocybe pelliculosa, Psilocybe semilanceata, Psilocybe serbica, Psilocybe stuntzii,* and many others. Low temperatures, even freezing, followed by warmer weather can stimulate flushes of many of these species. Lower temperatures may also stimulate primordia, which rest unless entering into this isotherm. Many temperate species will only fruit when the minimum daily temperature goes below 50°F (10°C) for several days.

For decades, I have been told that the Indigenous people of Mesoamerica disdained *Psilocybe cubensis* due to its association with the Spanish invasion, favoring other native species like *Psilocybe aztecorum, Psilocybe caerulescens, Psilocybe zapotecorum*, and *Psilocybe mexicana. Psilocybe cubensis* is known in Mexico as San Ysidro, the patron saint of agriculture. (San Isidro Labrador was born in Madrid, Spain, in 1802.)[6] However, since most of these species are found irregularly and depend on local weather, many of today's practicing shamans and guides are taking advantage of the ready supply of laboratory-grown psilocybin mushrooms like *Psilocybe cubensis*. This is an interesting example of in vitro propagation, a Western technological innovation, helping Indigenous traditions survive under the pressure of increasing demand and decreasing availability of sacred fungi. Indigenous traditions are never static; they constantly adapt over time by adopting or innovating new techniques. Psilocybin builds bridges between cultures as modern technology protects ancient practices.

Psilocybe semilanceata thrives in saturated pastures and near ponds. Liberty Caps tend to have short stems in short grass and long stems in tall grass.

This pond is in a lowland ecosystem that is a good habitat for *Psilocybe semilanceata*—especially when grazed and enriched by cows, sheep, and, to a lesser extent, horses.

This is me contemplating the meaning of a "No Mushroom Picking" sign. These signs alert passersby that mushrooms of interest are nearby. By reaching for a mushroom, photographing it, even touching it but technically not picking it, I thus remained fully compliant with the law here. However, one well-known mycologist I know was busted and fined for photographing a mushroom even though he was a professor of mycology at a university.

60 to 80°F (15.6 to 26.7°C): *Panaeolus cinctulus, Panaeolus cyanescens, Psilocybe ovoideocystidiata, Psilocybe zapotecorum, Psilocybe caerulescens, Psilocybe mexicana, Psilocybe cubensis*, and *Psilocybe stametsii*. Temperatures lower than 60°F (15.6°C) inhibit these species' growth, whereas temperatures above 80°F (26.7°C) cause many of them to mature faster but inhibit new primordia formation. This isotherm represents the range for fruitbody growth, not necessarily primordia formation.

Mushrooms grown outdoors are often much larger than those grown indoors on the same substrate, likely due to the contact with beneficial soil microbes, better airflow, and sunlight. If you add the mycelium to cow or chicken manure compost, the flush could be massive. However, making the right compost can be tricky. Use your imagination—there are lots of options for mycelial mileage (for more ideas, see the book *The Mushroom Cultivator*).

Although cow and sheep pasturelands are prime habitats for grassland species, several psilocybin mushrooms grow in grasslands not enriched with manure. In the Pacific Northwest, two examples stand out: both are the grassy meridians separating one-way traffic at the entrances

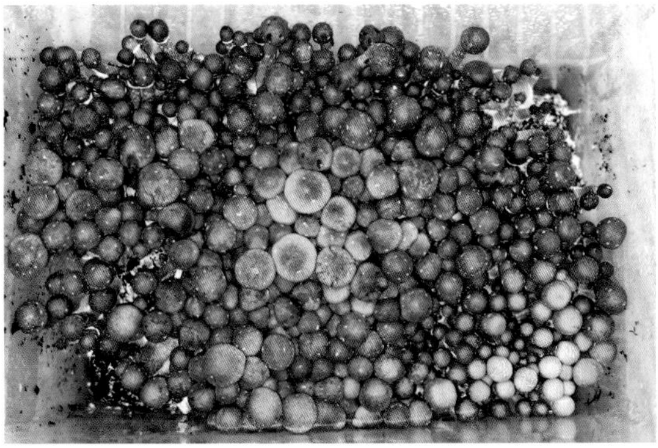

Psilocybe cubensis densely fruiting on a mixture of rice and sunflower-millet bird seed. Approximately one-third of the dry mass became dried mushrooms from the first flush. (I perfected this technique accidentally.) This psilocybin species is the most cultivated psilocybin mushroom in the world. Once it has fruited, the myceliated grain can be planted outdoors.

Psilocybe cubensis producing a second flush from the same mycelium featured in the above photograph, but this time outside in the garden a couple weeks later. This mycelium was covered with moist soil and watered for 1 to 2 days. Ten days later, this fruiting formed. (See my book *Growing Gourmet and Medicinal Mushrooms* for spawn-making methods and recipes.)

to University of British Columbia in Vancouver in British Columbia, Canada, and The Evergreen State College in Olympia, Washington. *Psilocybe semilanceata* has been prolific at these locations, and this species can favor traffic zones with high visibility. Soccer, rugby, football, and baseball fields as well as golf courses are also prime locations for psilocybin mushrooms. At Evergreen, I once witnessed the college soccer team suspend the game so psilocybin mushrooms would not be trampled underfoot, to the initial bewilderment of the other team, whose members soon enough joined in the fun of psilocybin mushroom picking.

Keep in mind, however, that herbicides and insecticides used in these manicured landscapes are a safety concern. Also, for the first decade after a lawn has been laid, it is not uncommon for classic wood-loving species of mushrooms to proliferate for years by decomposing the buried wood. Frequent watering stimulates growth. The appearance of wood-loving (lignicolous) psilocybin species in lawns and fields underscores that one cannot rely exclusively on signature habitats, especially newly constructed habitats, as a key indicator. With psilocybin mushrooms, exceptions rule.

Several species of lignicolous mushrooms grow in grassy areas with wood debris or underneath trees. This mixture of grasses and wood chips stimulates fruiting of *Psilocybe azurescens* (look for dunes with grasses), *Psilocybe aztecorum* (search under pines), and *Psilocybe ovoideocystidiata* (try along riparian zones with box elders). These interface environments, located between forests and grassy areas or in forests with grasses, are prime habitats for many psilocybin species.

A natural hardwood (alder) grove like this one supports wood chip–loving (lignicolous) psilocybin mushrooms. Lignicolous mushrooms like *Psilocybe azurescens* and *Psilocybe cyanescens*—to name a few of many—love grasses and decomposing thatch interspersed with fallen branches, forest detritus, and wood chips.

White Spores

Chapter 3

How to Identify Psilocybin Mushrooms

It's nearly impossible to accurately identify a mushroom species in the wild without having a good understanding of mushroom morphology, or form, and the various stages mushrooms undergo throughout their life cycle. While psilocybin mushrooms are generally similar to other types of mushrooms, there are a few disambiguating elements that set them apart.

Golden Top (*Psilocybe cubensis*) with a falling membranous partial veil forming the annulus. Once the veil begins to fall, it descends in a few minutes as the cap rapidly expands. Note the lines of spores on upper surface of falling veil.

Mushroom Morphology

More than 99 percent of all psilocybin mushrooms consist of a cap (pileus), gills (lamellae), and stem (stipe).[1] The cap protects the emerging mushroom from rain and wind, providing a still-air microclimate for the spore-bearing gills underneath the cap to ripen with spores. The stem elevates the mushroom cap and gills into the air, where the spores can freely float to new habitats.

Some of the psilocybin species, such as *Psilocybe cubensis*, also have a cobwebby (cortinate) or sheath-like (membranous) partial veil that extends from the cap edge to the stem to

David Sumerlin examines and identifies some of the mushrooms collected in one day from a group foray. Note mushrooms are sorted by spore color.

Azzies (*Psilocybe azurescens*) have a cortinate partial veil that may or may not form a fibrillose annular zone, soon decorated with dark purple brown spores. Notably, the annular zone is ephemeral—often disappearing as the cap expands.

protect the spore-producing gills (lamellae). The outer edge of this veil separates from the cap as the mushroom matures, gracefully descending over a few minutes and coinciding with the first release of the spores. The upper side of this membranous partial veil is often decorated with ridges of purplish brown spores. These lines of spores reflect the spaces between the gills above (see page 63).

When the veil has completely fallen, often the veil remnants create a ring (annulus) on the stem. In many mushrooms, a membranous partial veil creates a membranous annulus. However, some of these rings form only weakly or, rarely, not at all. The ring is often lost over time due to wind and rain. Keep in mind that every phenotype can be different. There are some weird phenotypes of *Psilocybe cubensis* that do not construct a good annulus at all.

In many cases, the type of annulus is important for accurate identification. One common type of partial veil is called a cortina. These are fine threads that stretch from the cap margin to the stem and typically break off as the mushroom matures, often leaving a fine ring of fibers on the stem, which also become adorned with spores. Such fibrillose veil remnants create a fibrillose annular zone or a cobwebby annulus. They, too, are fragile and can be lost in many specimens as the mushrooms mature and are exposed to weather. This is why it is important to have a full range of specimens, from youngest to oldest, for accurate identification.

Psilocybin mushrooms typically produce spores from basidia. In most species, each basidium produces four spores that may not all be compatible with the other spores. Most psilocybin mushrooms have two genes that control the mating compatibility of spores. One individual mushroom can produce spores with four possible mating types (like sexes).

The Mushroom Life Cycle

The mycelial stage represents more than 95 percent of the mushroom life cycle. Upon spore germination or mycelium translocation, the mycelium must establish a mycelial mat sufficient to generate the mushroom structure, called a fruitbody. This is a never-ending cycle from spores to germination to mycelium to mushroom fruitbodies to more sporulation. Spores can germinate immediately or rest latent for years. We do not fully understand why.

General forms of mushrooms, veils, and stems

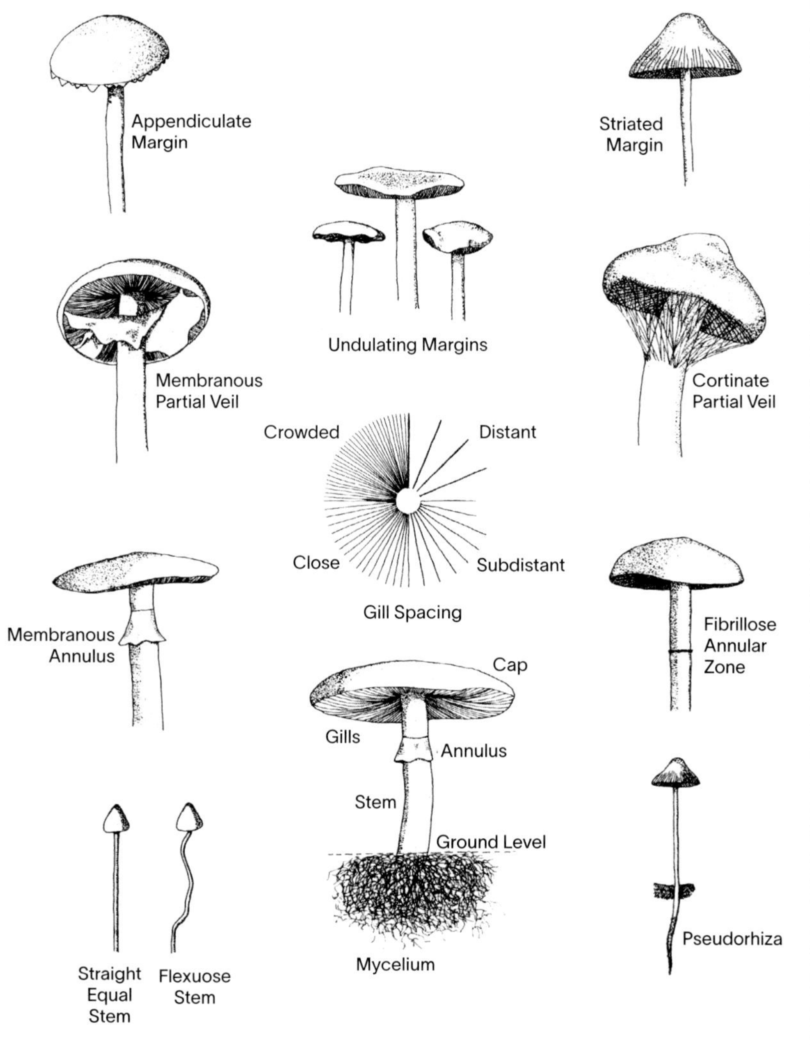

General forms of gills, spacing, and attachments

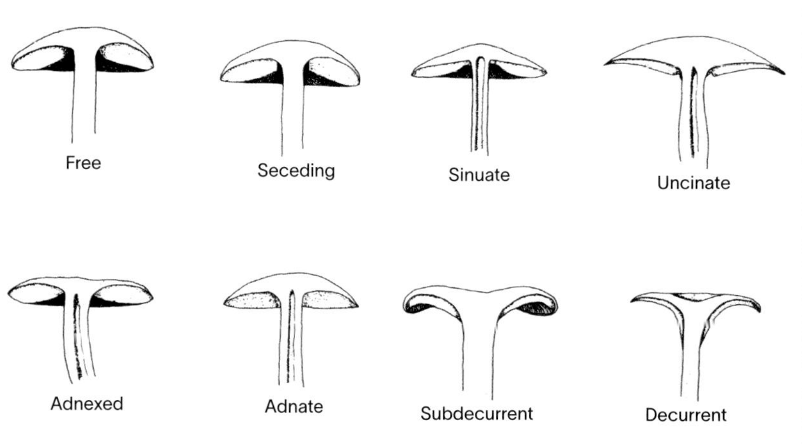

General forms of caps

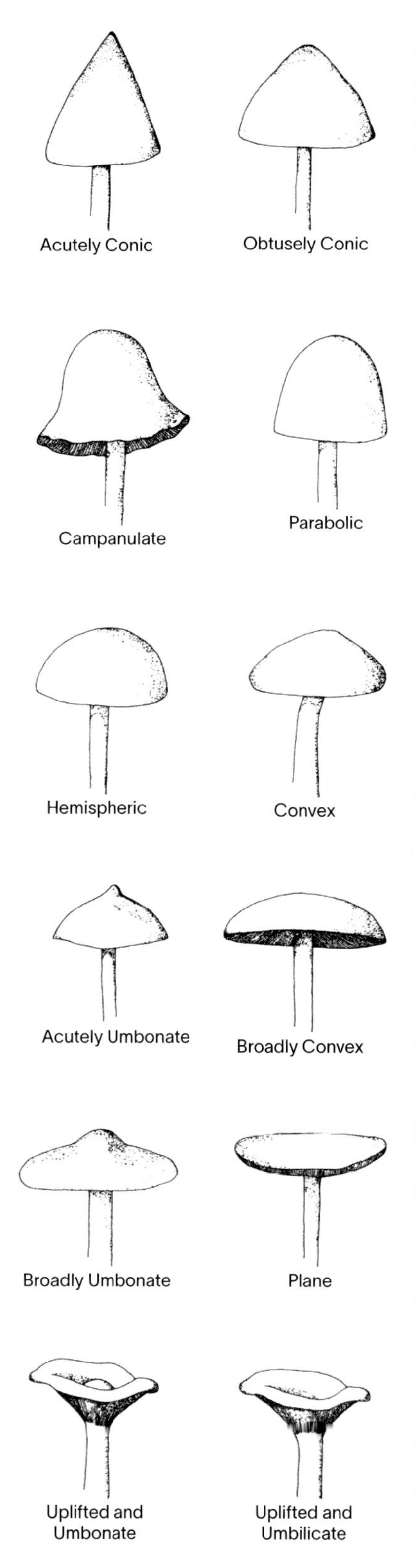

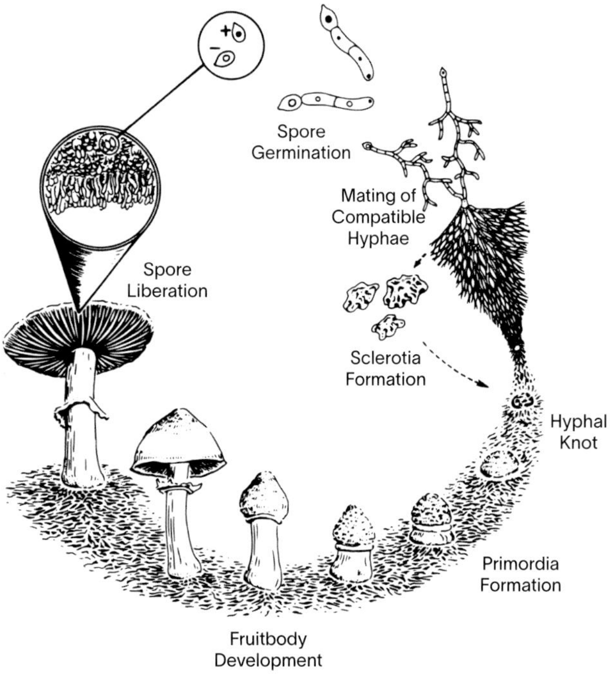

Above is the mushroom life cycle generally representing two groups of psilocybin mushrooms: *Psilocybe mexicana* and *Psilocybe tampanensis* versus *Psilocybe cubensis*. *Psilocybe mexicana* and *Psilocybe tampanensis* have evolved an alternative route for survival: the formation of a sclerotium, a hardened mass of resting tissue. The life cycle of *Psilocybe cubensis* lacks this feature. Sclerotia are stimulated to form when the mycelium is grown in darkness and can later fruit into mushrooms when exposed to light and water.[2] Sclerotia can survive many months before sprouting mushrooms.

The sclerotia of *Psilocybe mexicana* ready for sale in the Netherlands, where the material has been popularized as "magic truffles." These sclerotia, typically low in psilocybin, will sprout into more psilocybin-rich mushrooms when planted into soil.

Mushrooms generally form in response to eight environmental stimuli:

1. influx of water (rain, runoff, riparian zones, floods, drenching, sprinklers)
2. change in temperature (dropping in the fall, rising in spring)
3. exposure to fresh air when mycelium rises up to near the surface of its habitat, called the soil–air interface
4. exposure to light
5. physical disturbance
6. limitation of nutrients
7. challenge by another fungus
8. physical barrier

Cultivators of mushrooms use many of these factors strategically to induce mushroom fruitbodies to form from mushroom mycelium.[3]

In most climates, the advent of the rainy season coincides with cooler temperatures. When mycelium is near the soil surface, it "exhales" carbon dioxide and "inhales" oxygen. Most species of psilocybin mushroom are light-sensitive, so even when these other factors are at play, healthy mushrooms will not form unless exposed to light. Mushroom cultivators are skilled in using carbon dioxide levels to control the length of the stems and using light to influence the diameter of the caps. Cultivated enoki mushrooms, *Flammulina velutipes* and related species, have long stems due to low light and high ambient carbon dioxide levels. This combination coordinates growth of the mushroom fruitbody into the shape of mushrooms commonly bought in grocery stores.

The last group of factors are disturbance, limitation of resources, and challenges by competitors. Many mushroom species are stimulated to form fruitbodies by landslides, ground disturbance (construction, earthquakes), raking the substrate (a technique also used by cultivators), flooding, or the impact of your footsteps as you walk across the mycelial patches—pressure sensors underneath your feet! This may be why many psilocybin mushrooms (like the Trail Psilocybe, *Psilocybe pelliculosa*) most often occur alongside trails or rocky roads.

A more complicated stimulus is a challenge by a competing organism, such as another mushroom species. It is as if the mycelium knows that a new habitat cannot be conquered because an adversary is already resident. Therefore, it's time to fruit! However, some species such as Turkey Tail (*Trametes versicolor*) parasitize the mycelial networks of other species.

Another interesting—albeit less frequent—stimulus is lightning strikes. Observers and lore-keepers in Europe, Asia, and the Americas report mushrooms forming after lightning strikes. Experiments in Japan showed that pulses of 50,000 to 100,000 volts of electricity lasting 1/10,000,000th of a second can shock mycelium into mushroom formation.[4] Since lightning occurs all over the world, it is not surprising that many Indigenous peoples link mushroom formation with lightning strikes. This is another reason why we should not doubt Indigenous wisdom. Science often proves folklore to be factually accurate centuries later.

Lightning on the horizon forecasts impending rainfall. Mycelium is a network of sensitive fibers and so I have often wondered whether those fibers can sense the incoming rains. Lightning strikes, thunder follows, then rain. The sound of thunder on the horizon travels for miles as long wave frequencies. I believe that mycelium networks are like strings on a musical instrument and reverberate to thunder's sonic signature. Would mycelia then anticipate the impending rainfall and prepare to absorb water? It makes sense that not only lightning, but also thunder, and even drumming, could stimulate mycelial growth. If so, would the gathering of humans in celebration with music, dancing, and reverie awaken mycelial networks, and in doing so enhance the flow of nutrients to create mushrooms and nourish plants to help humans better survive?

Psilocybe mexicana mushrooms fruiting from sclerotia buried in soil.

Psilocybe pelliculosa loves to grow along forest trails and roadsides in the Pacific Northwest.

This idea seemed far-fetched to many people I have mentioned this to, but other mycologists have confirmed that vibrations and electricity stimulate mushroom growth.[5] Mycelium is sensing, if not literally listening. Fungi awaken, competitively, to every opportunity that increases their survival and ultimately ours.

The Rule for Identifying Most Psilocybin Mushrooms

Years ago, I summarized a simple rule for identifying psilocybin mushrooms and it still stands.

If a
gilled mushroom
has
purplish brown to black spores
and
bruises bluish,
then it is very *likely* a
psilocybin-containing mushroom.

Gilled mushrooms and purplish brown to black spores are relatively easy-to-understand features. Observing these two characteristics, plus bluing, narrows the field of candidates to primarily the psilocybin-active species. Bluing is a bruising reaction wherein the mushroom becomes bluish with handling or stress from repeatedly getting wet and drying. The mushrooms are not blue to begin with, like some *Stropharia* species, which are psilocybin-inactive (see photo on page 72). Psilocybin is a stable chemical; psilocin is not. Psilocybin is a pro-drug for psilocin, which means upon ingestion psilocybin is metabolized into psilocin, the active ingredient that changes your consciousness by docking with the 5HT2A and related neuroreceptors, particularly in your brain.[6]

Psilocybe azurescens, a purplish brown–spored mushroom, quickly bruises bluish with handling as psilocin is enzymatically degraded. This bluing reaction is typical of many potent psilocybin species.

When psilocybin mushrooms bruise bluish, it is the result of an enzymatic conversion of psilocin into compounds that are not psychoactive.[7] The phosphatase enzyme (PsiP) creates psilocin from psilocybin, which is then decomposed by the laccase enzyme into a heterogeneous mixture of quinoid psilocyl oligomers—compounds similar to indigo, hence the blue color. The strength of the bluing reaction in psilocybin mushrooms indicates how strong the mushrooms once were, meaning that they have lost potency. (The chemical process theoretically allows for bidirectionality back into psilocin.) If you find a group of psilocybin mushrooms that bruise strongly bluish, it is a good indicator that the whole colony is likely *very* potent. Outlier exceptions include the potent Liberty Cap (*Psilocybe semilanceata*), which barely bruises bluish, if at all, because of

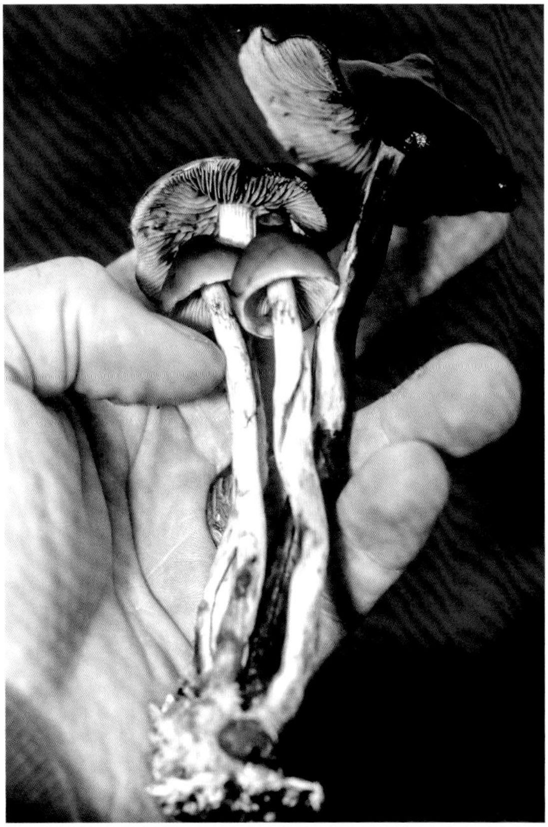

Psilocybe azurescens can become nearly black from bruising or freezing. I would not consume the heavily bluing mushroom on the right. It is rapidly decomposing.

Bluing reaction of *Psilocybe baeocystis* and *Psilocybe cyanofibrillosa*, two psilocybin-active, purplish brown–spored species.

Like many psilocybin-rich mushrooms, *Psilocybe cyanescens* turns blue the day after a deep frost when temperatures again rise above freezing. Once thawed, these blue mushrooms will quickly rot.

its high psilocybin and low psilocin content. And if this species does show bluing, it is typically in the fuzzy mycelium attached to the stem base. Most specimens of psilocybin-rich, psilocin-poor species lack bluing reactions but are still very potent *and* stable for long-term storage.

Exceptions to The Rule

There are some psilocybin-active species that have rusty brown spores and whose fruitbodies may or may not noticeably bruise bluish in all specimens. Some rusty brown–spored, poisonous species are look-alikes to these psilocybin species. This concerns me because misidentifications of rusty brown–spored mushrooms could have deadly consequences. These species must be memorized individually. On the other hand, some mushrooms are naturally bluish in color, but not from bruising. I know of no poisonous gilled mushrooms that are naturally bluish *and* also have purplish brown spores. This begs the question: Are the mushrooms you found naturally blue in color versus bruising blue? At times, this can be difficult to determine. The bluing reaction is a potential, but not definitive, feature that identifies a psilocybin mushroom. *Conocybula*

To create a very potent "blue juice" that can be stored for future use, place small pieces of *Psilocybe azurescens* (or many of the bluing *Psilocybe* species, including *Psilocybe cubensis*) in an empty jar, cover with ice, and refrigerate at 36–39°F (2–4°C) for 2 or 3 days. The resulting cold-water extract can also be frozen into blue cubes (to better preserve potency) or mixed 50:50 with ethanol (vodka) to create a liquid elixir. (The ethanol prevents souring.) This extraction method also works with cold water. Long exposure to hot water will turn the extract brown, although variably and often, but not always, de-potentizes the mushrooms. Because water temperature extraction is a delicate process, I find it safer to use cold water to create blue juice.

Blue juice made from the Wavy Cap, *Psilocybe cyanescens.*

Blue juice made from the Golden Top, *Psilocybe cubensis.*

cyanopus (rusty brown spored), *Inosperma calamistratum* (brown spored) and *Mycena amicta* (white spored) have distinct bluish tones at their stem bases—but if the stem base was not also collected, this critical identification feature would be lost. *Conocybula cyanopus* produces psilocybin, whereas *Inosperma calamistratum* and *Mycena amicta* do not. Another Mycena, *Mycena pura*, can be blue in color and is poisonous, containing the toxin muscarine. In the photos on pages 71 and 72, can you tell which one is psilocybin-active without reading the captions? If you used the bluish color at the base of the stems as your only criterion, you could mistakenly identify one as a psilocybin mushroom when it is not. This is why you must learn mushrooms species by species.

Notable exceptions to the The Rule apply to at least three members of the Strophariaceae family: the group that encompasses *Stropharia aeruginosa*, *Stropharia caerulea*, and *Stropharia cyanea*. They are related to the genus *Psilocybe* but do not contain any psilocybin. They can be brilliantly and beautifully blue all over, or sometimes spottily, but most often bluish at the stem's base. The edibility of this cluster of species is undetermined. Some anecdotal reports claim they cause gastrointestinal distress, so please do not consume them! They are gorgeous to look at, however.

Staying focused on purplish brown– to black-spored, *gilled* mushrooms that bruise bluish minimizes the risk of misidentification. Many mushroom species are bluish or can stain blue—especially the gilled species in the genera *Leptonia, Entoloma*, and *Stropharia* and in non-gilled species in the soft-pored family Boletaceae (e.g., *Xerocomellus chrysenteron,* formerly known as *Boletus chrysenteron*, and *Leccinum scabrum*) and the toothed Hydnaceae (*Hydnellum caeruleum* species complex). Even though a bluish color is one trait that *could* indicate psilocybin content, it is not the *only* one that should ever be used to conclusively identify a psilocybin species.

Mycena amicta is a white-spored, psilocybin-inactive mushroom. Note the blue tone at the base of the stem. *Mycena amicta* can display a range of colors.

Mycena pura, a white-spored mushroom, can be bluish toned but is poisonous, containing the toxin muscarine.

Inosperma calamistratum (= *Inocybe calamistrata*) is one of several closely related *Inocybe* species that turn blue but is psilocybin-inactive. Note the dark bluish tones at the base of the stem. Many Inocybes contain the poison muscarine.

Beware: Some Mushrooms with Rusty Brown Spores Can Be Deadly

When you stray from gilled mushrooms with dark purplish brown to black spores to those with rusty brown spores, you have entered the danger zone. Two genera of rusty brown–spored mushrooms, *Galerina* and *Pholiotina*, host some of the deadliest poisonous mushrooms known to exist, sharing toxins similar to those of the statelier *Amanita* species. *Galerina marginata* (Batsch) Kühner (= *Galerina autumnalis* Smith and Singer) and *Pholiotina rugosa* (Peck) Singer (= *Pholiotina filaris* (Fr.) Peck) can kill you with their potent liver toxins called cyclopeptides.

This group of rusty brown–spored mushrooms is further complicated because *Conocybula cyanopus*, a species formerly known as *Conocybe cyanopus* (G.F. Atk.) Kühner, a blue-footed psilocybin mushroom, was placed into the same genus as its deadly cousin, *Pholiotina rugosa*. The newer name is *Conocybula cyanopus* (G.F. Atk.) T. Bau & H.B. Song and they both share the same rusty brown spore color. Moreover,

Psilocybe cyanescens is a purplish brown–spored, psilocybin-active mushroom. Note the bluing on the stems.

Stropharia aeruginosa, a gorgeous, bluish-toned, purplish brown–spored mushroom species that is psilocybin-inactive. Edibility is questionable, unknown. (Caution is advised: Do Not Consume.)

Conocybula cyanopus (formerly called *Conocybe cyanopus* and *Pholiotina cyanopus*) is a rusty brown–spored, psilocybin-active mushroom. Note the bluing at the bases of the stems. If the stem bases are lost while picking, this species cannot be identified macroscopically.

The pinkish brown–spored *Entoloma serrulatum* (= *Leptonia serrulata*) group and related *Entoloma* species can be bluish but are psilocybin-inactive. The edibility of these species is unknown. (Caution is advised: Do Not Consume.)

there is a psilocybin-active but rare Galerina, called *Galerina steglichii* Besl., that was originally reported from Germany but is not yet known from North America. An unnamed bluing Galerina grows in Colombia, South America. It is only a matter of time before these wood chip–dwelling species take root in North America.

That two psilocybin mushrooms species can be classified within the same genera, Galerina and Pholiotina, as two deadly poisonous species underscores that a mistake in identification can be fatal. All four of these are rusty brown–spored species. All are typically small. All can co-occur in the same habitats as psilocybin species. An unfortunate, careless psilocybin mushroom–motivated forager could easily collect *Galerina marginata* and/or *Pholiotina rugosa* with deadly consequences. This has happened. People have died from mistaking a Galerina, thinking it was a Psilocybe.

Another genus hosting both psilocybin-active and poisonous species is *Inocybe*—a large genus of more than 1000 mostly mycorrhizal species. Mycorrhizal species grow in association with host trees in a symbiotic relationship, whereas the other genera containing psilocybin mushrooms are all saprophytes, or decomposers. *Inocybe* also contains toxic species that produce muscarine with potentially deadly consequences if consumed in sufficient quantities. Do some species coproduce psilocybin, muscarine, and/or other toxins? It's very possible, although not yet known. The chemistry of mushrooms has been barely explored. Focusing on the species with purplish brown to black spores avoids this potential hazard.

There are just a few deadly poisonous mushroom species. Many more toxic species are mildly poisonous. Typically, the effects of the most deadly poisonous mushrooms take several hours to be felt. However, poisoning from muscarine-producing mushrooms, which can be found among Clitocybes, Inocybes, Mycenas, and Omphalotus, can have an onset of symptoms in as little as 15 to 60 minutes. Anyone suffering from underlying illnesses could exhibit atypical symptoms, so if you feel ill, it's best to call the nearest poison control center, which in most cases can connect you with an expert. Make sure you save specimens for further analysis, if needed. And if you photograph the mushrooms suspected of causing illness, make sure you photograph the gills of the more mature mushrooms. Mycologists who work as consultants for poison control centers need to know the spore color of the questionable mushrooms. Although making a spore print is important, waiting overnight for a spore print may delay diagnosis and suggested treatment.

While this book is intended to be a guide to psilocybin mushrooms—not poisonous ones—the topic of poisonous mushrooms deserves your careful attention. I will not go into depth here, so I do recommend several books and online resources (see page 246). Please be certain of a mushroom's safety before consuming it or recommending it to others. Your life—and theirs—may depend on correct identification. We all share a responsibility to minimize the potential harm and maximize the benefits of magic mushrooms.

Pholiotina rugosa (= *Pholiotina filaris*) is a rusty brown–spored mushroom that can kill you. It can grow in a variety of habitats where soils are enriched with decomposing woody debris. Note the rusty brown spores on top of the annulus.

How to Make a Spore Print: It's Easy and Fun

Most psilocybin mushrooms have dark purplish brown to black spores. A minority of these species have rusty brown to salmon spores—but so do some deadly poisonous mushrooms. Out of an abundance of caution, newbies seeking to identify psilocybin mushrooms might want to focus their efforts on the species with purplish brown to black spores, which are also more abundant and easier to identify.

One of the most critical ways to avoid a poisonous species and find a psilocybin-active one is to properly identify the color of the mushroom's spores. Although eagle-eyed collectors can detect spore color in mushrooms growing in their natural habitats, a more certain method is to make a spore print.

Here's how to make a spore print:

1. Collect a range of adolescent to mature mushrooms when the caps are nearly fully expanded.

2. Sever the uppermost part of the stems from the caps and place the caps, gills down, on a piece of paper. White paper is the best choice most of the time. However, should the mushrooms not have dark spores or have white spores, colored paper or a combination of both white and colored paper will allow you to better detect spore color.

3. Place a bowl over the mushroom caps on the paper to keep the spores from blowing away; don't let the bowl press down on the caps. Let them sit overnight. Usually by the next day, the spores will have fallen following the radiating symmetry of the gills. If the candidate psilocybin mushrooms are small, the caps will shrink due to evaporation.

4. Uncover the mushrooms that rested overnight and assess the spore color. Sometimes, if the mushrooms are too old or dried out, they won't release spores. If this happens, try again with younger, fresher specimens.

Starting spore prints of the Wavy Cap, *Psilocybe cyanescens* (the day after collecting the fresh mushrooms).

Spore print of *Psilocybe cyanescens* (the second day after collecting the fresh mushrooms). The Wavy Cap gives a variable spore print due to the undulating cap margins.

Here are some examples of good spore prints from fresh mushrooms. Mushrooms that are drying out do not produce good spore prints. (Several of my books, including *Growing Gourmet and Medicinal Mushrooms* and *The Mushroom Cultivator*, have complete instructions for generating cultures from spores under sterile conditions that you can create at home.)

As shown in these photographs to the right, *Psilocybe allenii* and *Psilocybe cubensis* spores become visible overnight—or in a few hours, depending on the mushroom. Covering the specimens with a bowl helps to keep the spores from blowing away. This purple brown to purple black to black spore color is typical for these species. The heavier the accumulation of spores, the darker the spore prints.

Infusing spores into the oil used to lubricate chainsaws is an innovative way to spread and protect woodland mushroom species. I added *Psilocybe azurescens* spores to canola (*Brassica napus*) oil. The wood faces of the fresh cuts and the chips created become inoculated as the trees and branches are trimmed. I gave this spored oil to Ken Kesey, who was greatly amused by the concept—and put it to good use on his land in Oregon. Spores could also be added to sprayers or sprinkler systems to inoculate new habitats. Join the spore liberation movement. Spread spores!

Spore print of *Psilocybe allenii*. Note how the caps have shrunk overnight.

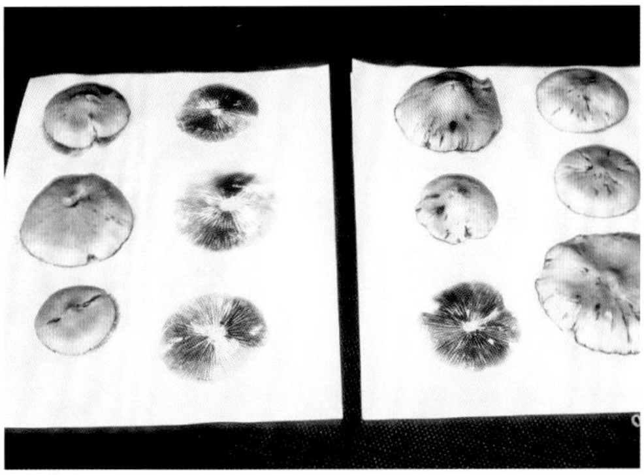

The spores of *Psilocybe cubensis* appear purple black to black en masse.

By infusing spores of *Psilocybe azurescens* in biodegradable chainsaw oil, woodcutters can spread populations of this and other mushroom species. You can take a spore print, scrape the spores from the paper or immerse the spore paper in the oil, and pour off the oil into the chainsaw reservoir, lubricating and sporulating the chain.

Chapter 4

How to Create a Psilocybin Mushroom Patch

Mushroom hunting is often a hit-or-miss exercise during which you expend a lot of energy for mixed or no results. One of the joys of life is to naturalize wild psilocybin mushrooms in a patch in your backyard—or outside of your high-rise apartment, if you have the space. Because of the risks of seeking what may be an illegal substance on property owned by others, the increasing popularity of mushroom hunting, and the resulting pressure on psilocybin mushroom populations, creating your own patch is also one small thing you can do to help protect psilocybin mushrooms and yourself. Among the advantages of having your own patch is getting to watch them grow. This intimate daily contact becomes a partnership. With your nurturing, they can reward you with their fruitful bounties for years to come, and you can reward them by keeping their mycelium running. It is fun to watch them show up each year, predictably, with minimal effort.

If you find psilocybin mushrooms on a field trip, many wood-decomposing species can be brought home and, in most cases, easily cultivated. Some of the species that are the easiest to naturalize are *Psilocybe allenii, Psilocybe*

The rhizomorphs at the stem base of many wood-rotting mushrooms can be clipped from the mushrooms and replanted into fresh hardwood chips to establish your own mushroom patch. The mushrooms featured here are *Psilocybe azurescens.*

Raking wood chips to prepare a mushroom bed.

Psilocybe azurescens, with its fuzzy stem butts and radiating rhizomorphs, is a species that can be easily naturalized. Note the bluish tones. Many psilocybin mushrooms have bluish tones at their stem bases, especially associated with the sensitive aerial mycelium.

aztecorum, Psilocybe azurescens, Psilocybe caerulipes, Psilocybe caerulescens, Psilocybe cyanescens, Psilocybe hopii, Psilocybe mescaleroensis, Psilocybe ovoideocystidiata, Psilocybe serbica, Psilocybe subaeruginascens, Psilocybe subaeruginosa, and similar wood-decomposing psilocybin species.

Wood chip–loving species producing luxurious rhizomorphic (stringy) mycelium, adorning the base of the stems, are generally easiest to naturalize to create new mushroom patches. The coprophilic (dung-dwelling) and pastoral species, on the other hand, are more difficult to naturalize. Dung decomposes rapidly, so the window of opportunity is shorter. And a few grassland species are especially complex to cultivate due to the not-yet-understood associations with grasses, their rhizomes, and the microbiomes within the rhizospheres they establish.

Wood chips take a few years to decompose before the cultivated mushrooms no longer form. The first year usually has the best fruiting, and then it wanes in the second and third years unless you annually replenish the patch with fresh wood chips. One alert British biologist noted a huge flush of *Psilocybe cyanescens* in the fall of 2000 after a turnaround was mulched the year before; the fruiting was estimated to be at least 100,000 mushrooms or more.[1] And I once witnessed a fruiting of *Psilocybe stuntzii* that boggled my mind—an eruption of side-by-side psilocybin mushrooms that literally elevated the landscape to a higher plane—directly across from a police station on Boat Street near the University of Washington, Seattle.

This happy mushroomer has collected *Psilocybe cyanescens*, trimming the stem butts from the mushrooms while picking. The stem butts can be planted—like tulip bulbs—into piles of fresh hardwood chips to generate mother beds of naturalized spawn, which can be later transplanted into a much larger mass of wood chips, expanding the bed exponentially.

Chipping small-diameter alder trees.

There are three primary methods of naturalizing mycelium:

1. inoculating wood chips (fermented or fresh) with laboratory-grown mycelium (aka spawn)
2. inoculating wood chips with stem butts of *Psilocybe* mushrooms
3. transplanting lenses of wild or naturalized mycelium

Wood chip–loving *Psilocybe* mushrooms, as well as the dung-loving Psilocybes typically have fuzzy stem butts that, when clipped, can be used as spawn. The stem butts magically transform, degenerating from solid fragments of tissue and regenerating into vigorously growing, diverging strands of luxuriously rhizomorphic mycelium. Curiously, the regions above the stem butts will rot with molds, while the stem butts attached to the mycelium that birthed them generally do not. The wood chip–loving psilocybin mushrooms are simpler to grow than the coprophilic *Psilocybe* species like *Psilocybe cubensis*.

I describe in detail how to create spawn in my book *Growing Gourmet and Medicinal Mushrooms* and how to inoculate from stem butts and transplant wild mycelium in my book *Mycelium Running: How Mushrooms Can Help Save the World*. But the following method is more direct, because you do not need a laboratory to grow spawn. It's also greatly improved by fermenting the wood chips for a few weeks before inoculation and using the resulting mother mycelium for further, massive, exponential growth.

Your backyard mushroom beds must be replenished with fresh wood chips each year, just like in nature when winter storms break off branches, feeding the wild mycelium that will bear fruitbodies the following fall. Mycelium hungers to be fed. Without replenishment with wood chips or newly created fermented and mycelium-colonized wood chips, your bed of wood chips will decompose into new, rich soil. I have repeatedly seen 6 to 12 inches (15 to 30 cm) of fresh wood chips decompose into 1 to 2 inches (2.5 to 5 cm) of dark soil in 2 years. Wood-based mushroom patches usually die out in 2 to 3 years unless annually refueled with fresh wood chips.

A surprisingly simple and successful replenishment method to expand healthy mycelium is to first ferment fresh wood chips in fresh water for a few weeks prior to inoculation. This low-tech method creates a complex microbial soup highly conducive for rapid mycelial growth. The wood chips I prefer come from hardwoods, notably alder, aspen, cottonwood, birch, oak, boxelder,

Ideal size of wood chips for making outdoor mushroom beds. Larger chips interspersed with smaller ones create fractally complex microcosms in which the mycelium thrives.

elms, and the like. I have successfully experimented with mixed woods, especially Douglas fir, sometimes with cedar mixed in. Should you use cedar, redwood, or juniper, it must make up a very low percentage of the total wood due to these woods' antifungal properties, which would inhibit the mycelium's growth. Straw could also be used, but it is harder to work with and the fermentation is not as successful as that produced with wood chips. Moreover, straw-grown mycelium has a much shorter lifespan than that growing in wood chips, which are more resilient and longer lasting.

Submerged Fermentation Method

This is my preferred and most excellent method for creating vigorously growing mycelium with enhanced immunity against competitors. I stumbled on this method by chance when I left birch-wood dowels submerged in a 5-gallon bucket for several weeks, during which a slimy film formed on top. When I drained the bucket of the smelly liquid and inoculated the dowels hours later, the leap-off of mycelium that ensued was impressive: fast, ropey, forking, and three-dimensional. Frankly, I was blown away. I then began to experiment with larger containers and wood chips and straw. Although I independently made this discovery, I soon learned that others had discovered this method as well, especially in Asia. (One of my mottos: "Good ideas recommend themselves.") I am gratified that others have also proven this method to be successful and am sure that others still can improve on it. The mycelium is the best teacher. Encourage its growth to form mushrooms—and we will follow.

First, it is important to know that, like us, mycelium has an immune system. When a pure culture of mycelium is grown in a clean-room laboratory that is free of other organisms, the mycelium's immune system is at a resting state. So, when you spread spawn grown from a pure culture into a bed of unsterilized wood chips, the mycelium struggles to stave off resident microbial competitors. This mycelium is what I call "immunologically naive." Once it survives its contact with microbes and adapts and grows, the mycelium becomes "immunologically educated."

Submerging wood chips in water and allowing them to ferment for several weeks causes the stagnant-water microbial ecosystem to quickly transition into anaerobic fermentation. I was challenged on this claim by a graduate student, so we did a DNA analysis of the fermentation broth. Indeed, the dominant bacterial populations went from aerobic species to anaerobic species by three orders of magnitude after several weeks.

Be forewarned that *Klebsiella* and other potentially dangerous bacteria can grow in fermented wood chips. Exposure to concentrated bacteria could be dangerous, particularly to those who are immunocompromised—and especially if you have lacerations. To minimize risk, wear gloves and do not touch your face, nose, or eyes after making contact. Wash your hands and gloves afterward.

Upon draining the water from the fermented wood chips, oxygen in the air is toxic to the anaerobes, hindering or killing them. The anaerobes become nutritious food for the mycelium, and the wood chips are tenderized. This fermentation method, therefore, gives immunologically naive mycelium an advantage by turning the tables in the mycelium's favor.

Mycelium, like us, is aerobic: it "breathes," inhaling oxygen and exhaling carbon dioxide. Since oxygen is toxic to anaerobes, their vitality

Wood chips, after being placed into totes, are submerged in water. The submerged ecosystem tends to become dominant with anaerobic bacteria. After 4 weeks, the fermented wood chips are drained and oxygen neutralizes the anaerobic bacteria. Then it's time to inoculate. You can use stem butts, transplanted lenses of naturalized mycelium, or laboratory-grown grain or sawdust spawn. All work well! Inoculation rates vary from 5 to 10 stem butts or 1 to 20 percent by volume of spawn to wood chips in a typical tote depicted here (27 gallons or 102 L).

Closeup of *Psilocybe cyanescens* myceliated wood chips. This vigorous mycelium is "immunologically educated" and at an ideal stage for transplantation or for use to expand into more wood chips at a recommended rate of 5 to 20 times.

From 8 to 12 weeks post inoculation, sooner in temperate climates and later if you have a deep-freezing winter, the wood chips are myceliated. This mycelium has a fortified immune system, as it has gobbled up microbes and created a mycobiome. Now it is ready to either be expanded into newly fermented wood chips (1 to 20×) or laid outside to create a bed.

Upside down block of myceliated wood chips 12-plus weeks after inoculation. Note that 95 percent of this wood chip block is solidly infused with and held together by mycelium.

declines as the mycelium is introduced and surges to gobble them up. I hypothesize that this sudden contact epigenetically activates gene expressions that stimulate mycelial growth. This form of new mycelium is no longer immunologically naive. It is microbially empowered and grows vigorously.

Similarly, when you transplant stem butts of mushrooms adorned with rhizomorphs or wild patches of mycelium, the mycelium typically takes off with rapid growth. It already knows how to mount immune defenses against many potential pathogens. Moreover, the mycelium favors mycelial-enhancing microbes, which create a defensive quorum. The mycelium's immune system is activated as it prebiotically favors guilds

Raking out myceliated wood chips into a 1- to 3-inch layer to create a mushroom bed.

Fruiting of mushrooms in late September from raked bed 2 to 3 months earlier.

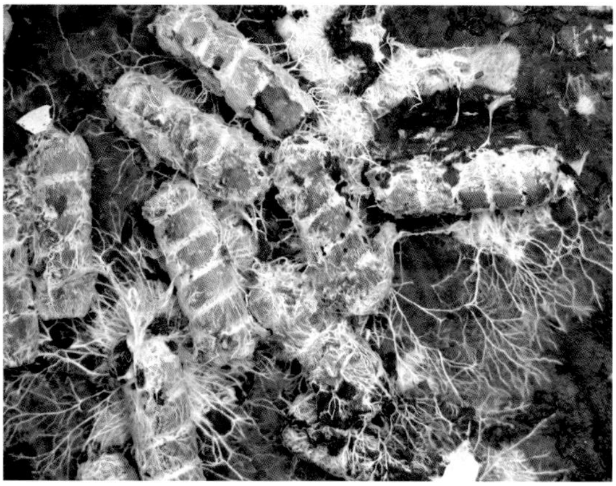

Instead of using only wood chips, you can soak hardwood (birch, oak) dowels using the same fermentation method. With both types of wood, lay them out in a shady place and top-dress with a shallow 2- to 4-inch (5- to 10-cm) layer of fresh wood chips. Or you can inoculate freshly cut hardwood logs to create mycelial rafts. These are but two of many ways you can keep the mycelium running.

Myceliated wood of *Psilocybe azurescens* species has a distinct fragrance. Note the tenacity of the mycelial rhizomorphs grasping the dowels.

of bacteria beneficial to its own survival, staving off competitors. This immunologically educated mycelium forms a community of microbial allies, a complex mycoshield of defense.

Once this mycobiome collective is established, the combined host defenses of its members synergize to allow the mycelium to enlarge exponentially. I am continually amazed at the vitality of mycelium in this condition. About 9 square feet (0.84 m^2) of mycelium can have trillions of crossings, end branchings, and apical tips. At each nexus of crossings and with each diverging tip, the mycelium gathers information that helps educate its immune system. Much like how a vaccination works, the mycelium may become increasingly immune to pathogenic threats. Mother mycelium at this state is treasured by those of us who cultivate; it retains strategic defenses and talents. Once launched into a supportive substrate, these rhizomorphs have the capacity to project downstream mycelium far and wide. Long live rhizomorphs!

A cluster of rhizomorphs of *Psilocybe azurescens* can hold tens of thousands of times its weight in wood chips. Mycelium prevents erosion, retains water, and holds soils together.

Creating Your Own Mushroom Patch

Using the fermentation method, these are the steps for creating an outdoor mushroom patch:

Step 1. Preferably using a wood-chipper, but by whatever means available, cut leaf-free hardwood branches into ¼-inch (6 mm) to 2-inch (5 cm) or slightly larger segments. I like to start chipping in the fall and winter, after the leaves have fallen and especially if wood chip–loving psilocybin mushrooms are fruiting so I can use their stem butts for inoculation. Spring chipping also works, but the incubation-to-fruiting window is shortened.

Step 2. Fill a lidded 10- to 27-gallon (38- to 102-L) container with the wood chips to a depth of no more than 18 inches (46 cm). Add nonchlorinated water until the wood chips are submerged. Cover with the lid, and let the container sit for 2 to 4 weeks outside, or, during winter months, in a cold, above-freezing space. Many types of containers can be used for the wood chip fermentation method. I like the common totes you can buy from hardware stores and garden centers, as shown on page 81.

Step 3. After several weeks, put on gloves and drain the water back into the earth. Do not let the water contact your skin, cuts, or other sensitive areas. Out of an abundance of caution, do not touch your eyes, nose, mouth, or any exposed cuts. This is especially important if you are immunocompromised. Make sure the drained water does not enter a river, lake, ocean, or the water table that sources your local drinking water. (I simply pour it back into the forest soil, with no apparent ill effects to the trees being exposed/nourished.)

Step 4. While keeping the fermented wood chips inside the container, spread them out so they are exposed to the air. You may have to aerate the wood chips occasionally by stirring them with a shovel, rake, or pitchfork. If you're using 5-gallon (19-L) buckets, simply pour the wood chips from one bucket into another and lay the bucket at an angle to maximize aeration. I always leave about 20 percent of headspace above the chips so they can out-gas. You may also want to add drain holes to the lower regions of the container, so the substrate can be colonized. Allow the container to sit for a few hours or overnight. There are no hard-and-fast rules here. Experiment to see what works best for you and your mycelium. This is a partnership. You and your mycelium both learn how to work together.

Step 5. Within the container, inoculate the fermented wood chips with laboratory-grown spawn, stem butts, or transplanted wild mycelium by simply mixing them throughout the new substrate. You could also use cardboard spawn, which I discuss in detail in my book *Mycelium Running*, or transfer lenses of naturalized mycelium. Your inoculation rate (i.e., the volume of your spawn, stem butts, or transplanted mycelium to the total volume of the wood chip substrate) can be as little as 1 percent and up to 50 percent; more than that is unnecessary. If using stem butts, I recommend five to ten stem butts per 4 gallons (15 L) of substrate in a 5-gallon (19-L) bucket; 10 percent is a good inoculation rate if using mycelium. The higher the inoculation rate, the faster the growth, and the sooner you need to further expand the mycelium. The key concept here is to keep the mycelium running!

Step 6. Cover your container loosely to prevent rain accumulation while allowing slow transpiration and evaporation. I place torn-up cardboard on top of the wood chips and store the container in the shade. The cardboard may also become colonized with mycelium as it reaches up. You can use this cardboard for further transplantation.

Step 7. After 4 to 8 weeks, given the many variables at play, you *should* see massive mycelial growth (see top right photo on page 81). Once overwintered either outdoors or preferably in a cold place above freezing, this mycelium is ready to be expanded into a new outdoor mushroom patch. At this stage of invigoration, the mycelium can be used as is or to inoculate 1 to 100 times its volume! (I recommend 4 to 20 times.) Note that the best time to plant your outdoor mushroom bed is by early spring for fruitings to occur in the fall. If you don't move the mycelium that has incubated in the totes, mushrooms will often form inside the container, but the fruitings are much fewer than if you had launched the mycelium outdoors in the creation of your sacred mushroom patch.

Find a spot with full shade. If your backyard or garden has exposure to direct sun, dry winds, or is otherwise in a less hospitable environment (e.g., with widely fluctuating temperatures), place a breathable cloth over the incubating bed or scatter pieces of cardboard on top of the wood chips to help protect the site where the mushrooms flourish. Placing your mother mushroom bed in a garden is ideal, in my opinion, because most of these saprophytes are friendly to vegetables, create great soil, will have easy access to water, and you can easily see them when they fruit. I have missed many fruitings by not being in the right place at the right time. Since you now have amassed vigorous, immunized mycelium, you can directly inoculate fresh, nonfermented wood chips. Spread newly chipped wood to a depth of 1 to 2 inches (2.5 to 5 cm), then spread your myceliated wood chips evenly across. Spread another 1- to 2-inch layer of fresh wood chips on top. This creates a horizontal mycelial plane in the middle of two layers of wood chips. Alternatively, you could mix the mycelium through the substrate, which is a common method to inoculate sterilized or pasteurized substrates. However, I find that using this planar inoculation method better guarantees success. Rather than having individual points of inoculation, each of which must struggle to overcome surrounding competitors, the contiguous sheath of mycelium creates a stronger, contiguous colony.

Typically, it takes at least 2 months for naturalized spawn to colonize fresh wood chips before the fruiting season—with the onset of conducive temperatures and rains. In the northern latitudes for most species, fruitings happen in early fall. During the grow-out season from spring through summer, the greatest concern is that the mycelium could dry out. Although mycelium releases water as it digests the wood chips, exposure to direct sunlight is harmful. Indirect or dappled sunlight is not too harmful but may delay early fruitings if the surface is highly exposed.

The *Psilocybe cyanescens* bed pictured on page 88, lower right, fruited more than a month later than the ones that have grass growing in them. This method requires an annual topdressing of fresh wood chips, but ultimately other fungi will take over. After 3 years of topdressing, I advise that you create a fresh patch using stem butts or new mycelium (from stem butts or spawn). Also, note that many spontaneous satellite colonies can sprout up unexpectedly across your landscape from spore fall. Always fun!

Step 8. Once you have laid down your layers of wood chips, disperse a small amount of grass seed on top to help mushroom formation later if the adjacent area is not grassy. I have a hunch that primordia form best in dew drops, especially as water droplets stream from the tips to the bases of grasses to settle on surface mycelium. Grass growth helps to create the high-humidity microenvironments conducive to mushroom formation. If your patch is located along a grassy area, the rhizomes will likely reach underneath and spontaneously create a grassy ecosystem on top of the chips.

Psilocybe cyanescens fruiting from fermented mycelium inside its incubation tote. If the mycelium is not transplanted and expanded, mushrooms will form within the species' preferred isotherm (40–60°F, 4.4–15.6°C). This same spawn could be expanded from 1 to 20 or even up to 100 times (!) on fresh, nonfermented wood chips. (I recommend 4 to 20 times.) Spawn run temperatures can be higher or lower than the isotherms conducive to fruiting. Mycelium has a high tolerance for temperature fluctuation; its fruitings do not.

A *Psilocybe cyanescens* bed incubating after inoculation in the spring. This breathable cloth allows air exchange and rain to drip through, while providing shade and keeping an incubation environment moist and conducive to mycelial growth. Grass grows underneath, which helps early primordia formation. This cloth is pulled off once the fall rains begin. When the grass is left to grow, the mushrooms become taller and are much more substantive than if the grass is removed. It is up to you to refine the best practice for maximizing yields. Importantly, allowing the grass to grow results in fruitings up to 1 to 2 months earlier than exposed beds and is my preferred method. Plus, the mushrooms are more incognito, in stealth mode, evading easy detection from potential mushroom marauders.

Step 9. Beginning in early September, if there is no rain, water the bed by spraying it for a few minutes every day or every other day to encourage the mycelium to reach up to the surface.

Step 10. In late September through early December in most locations in the northern hemisphere, mushrooms will flourish, giving you an abundance of not only mushrooms but also stem butts that are ideal for next year's inoculations.[2] It's best to harvest your mushrooms in their adolescent stage before the caps flatten and heavily sporulate. If you are eager to collect spores, harvest mushrooms that are mature. If they become too mature to harvest, no worries, because their spore cast can help invigorate the underlying mycelium and also create more satellite colonies near and far. Once spores are released, the growth becomes exponential.

Step 11. After harvesting your mushrooms, it is an ideal time to chip more leaf-free wood. If you use fresh wood chips, top-dress your bed ideally with no more than 1 to 2 inches (2.5 to 5 cm). I have found 3 inches (7 to 8 cm) is the maximum recommended thickness. Bigger chips allow more respiration than smaller chips. Layering chips any deeper than this could send the bed into a state of suspended mycelial hibernation and encourages competitor fungi. The new wood-chip layer needs to be shallow and replenished annually.

If you decide to expand your patch at the time of fruiting, you can ferment more wood chips for at least 2 weeks and inoculate with fresh stem butts from the mushrooms you just harvested, or scoop up myceliated wood chips (naturalized spawn) to inoculate and let them incubate over the winter. The mycelium of this northern species

This same bed as in the top image produced several massive flushes of mushrooms 5 months later, but weeks prior to beds inoculated at the same time that did not have a cloth cover and were instead exposed to the open air. Moreover, successive fruitings in the grassy bed continued for nearly 3 months—and for every week thereafter during the fall. This bed had the advantage of scarred logs embedded into the wood chips. Hence, it is a better platform for fruiting.

The Wavy Cap, *Psilocybe cyanescens,* tends to fruit in clusters like this young grouping.

Psilocybe cyanescens from primordia to maturity.

can freeze solid and regrow. Here, the strategy of benign neglect works. Basically, leave the mycelium alone; just check on it every few months by carefully digging into the wood chips. You may not see much growth in the winter, but by midspring the mycelium surges again.

Extending the Lifespan of Your Mushroom Beds

Since psilocybin mushrooms are saprophytes (aka saprobes; they live on dead or decaying matter), once they have decomposed fresh wood chips, they move on. In a practical sense, the mushroom patch dies out, leaving rich soil behind. I modified the shiitake log method for prolonging the mushroom patch by inoculating scarred alder logs with direct mycelial contact. Analogous to how quickly a stove will burn through wood chips compared to a log, the mycelium takes years, rather than months, to decompose an alder log.

The method is quite simple. Once you have established a bed luxuriant with mycelium, cut young hardwood (in this case, alder) trees and scar the sides (i.e., remove sections of bark so the interior wood is exposed) with a hatchet, machete, or chainsaw. Place the logs in parallel directly in contact with the mycelium to allow direct colonization into the log. These mycelial rafts may prolong the lifespan of a *Psilocybe cyanescens* bed, for instance, from 1 or 2 years to many more, provided a thin annual topdressing of newly cut wood chips is applied. Plus, you can add new scarred logs in between the logs from the previous year so the mycelium jumps into those. Once colonized, the logs can be moved to a new location for creating satellite colonies—and the inoculated logs make great gifts for friends and family.

Once the scarred logs are placed into contact with myceliated wood chips, the mycelium colonizes the wood. The log on page 88 shows mycelial growth about 3 months after contact.

Harvesting and Storing Psilocybin Mushrooms

The best time to harvest psilocybin mushrooms, in my opinion, is before or at the earliest stages of sporulation. This is typically when the mushrooms are mid-adolescent, meaning the gills are not yet producing many spores. Spores have no psilocybin. This is one of the reasons that the sale of spores has avoided being legally controlled on the federal level in the United States. (Although some states, like California, restrict the sale of psilocybin mushroom spores.)

As the fruitbody forms and emerges, its flesh is initially very dense. Over time, the density of the mushroom flesh reduces as the gills mature. Although it has not yet been proven, in my experience the potency is maximized when the fruitbodies are still in what I call the drumstick stage with a tight and small veil along the cap margin. In a matter of hours, sometimes minutes, the caps expand, the veil begins to fall from the outer cap edge, a fibrillose or membranous ring (annulus) can form, with spores soon discharging en masse. As psilocybin mushrooms' dense tissues transform into gills and spores, there is a corresponding drop off in tissue mass containing psilocybin and psilocin. From a practical point of view, it is difficult to find mushrooms at this exact state when collecting in nature. Most discoveries are serendipitous. If you are a grower, however, you can watch your mushrooms develop by the minute. The acceleration of their rate of growth from a thumbnail-sized baby mushroom to the closed-veil drumstick stage to adolescence can occur within a few hours in warm-weather species and days longer in cold-weather species. *Psilocybe cyanescens* and its allies are usually slower to sporulate, waiting until the caps expand to broadly convex.

Using a machete, scar the logs to create entry wounds for the mycelium to infiltrate into the interior wood. This method creates a reservoir of mycelium that lasts longer than in wood chips. Scarring the logs can be done in the fall to early spring. I prefer the fall after the last flush to allow best colonization. The myceliated logs can be moved to another bed of uninoculated wood chips as a central axis for more mushroom patches.

The mycelium can jump onto an alder log laid in direct contact. This augments the fruiting for next year.

An outdoor bed near water benefits from higher humidity.

Psilocybe cyanescens mushrooms forming in beds fortified with scarred logs and infused with grass produces very large fruitbodies. Note their robustness and bluing on the stems.

Psilocybe cubensis has a membranous partial veil that falls as the caps expand, synchronous with sporulation. *Psilocybe cubensis* is best harvested at the beginning of the veil's tearing. Spores are released in small quantities at this stage and will increase exponentially within a few hours. The purple brown spores adorn the upper surfaces of the membranous rings (annuli). Younger specimens are preferred; bigger is not better. Spore mass can be substantial; in oyster mushrooms, for example, the spores generated can be more than 20 percent of the fruitbody. With *Psilocybe cubensis*, I would expect around 10 percent of the mycomass to be invested in spore formation. A common misconception about maturing mushrooms is that larger mushrooms are heavier. In fact, they are less dense and have less mass than at their adolescent stage. The mass is highest prior to the partial veils falling, the caps flaring, and spores being released by the millions.

I emphasize harvesting when the veil on the cap is closed for five reasons:

1. The potency of the mushrooms is optimized at this morphology.
2. Once mushrooms sporulate, they attract insects, molds, and bacteria and rot quickly.
3. Some people are allergic to mushroom spores, and some to the spores of *Psilocybe cubensis* more than those of other mushroom species.[3]
4. Spore mass comes from a substantial proportion of total mushroom mass, and the conversion of mycomass to spores reduces the potency.
5. Spore mass laid on the surrounding mycelium may prevent more mushrooms from emerging, not only physically but also because numerous incompatible strains can form, competing with the underlying mycelium and also, in some cases, investing in hyphal mating via anastomosis.

Preserving Psilocybin Mushrooms and Their Potency

Once harvested, psilocybin mushrooms can be dried in a food dehydrator set at a low temperature or, better yet, air dried at room temperature downwind of fans blowing dry air over them. Vacuum packaging dried mushrooms in airtight bags and then freezing them is also a preferred method. The drying temperature should be 75°F (24°C) or below. Drier air will increase evaporation rates. Smaller mushrooms dry more quickly than larger ones and, hence, may retain higher potency. Once dried, freezing has generally

Psilocybe cubensis harvested at the ideal stage: when the veils are closed and before the gills are massively sporulating.

been considered the best way to store psilocybin mushrooms, although there are some reports that dried mushrooms stored at room temperature, in darkness, hold their potency for a long time. Mushrooms higher in psilocybin and lower in psilocin, which is less stable, tend to retain their psychoactivity longer. Many variables are at play, however.

All psilocybin mushrooms lose potency over time. For instance, dried *Psilocybe cubensis* typically has 1 to 2 percent, rarely up to 5 percent, psilocybin/psilocin content, which is equivalent to 10 to 40 mg of psilocybin/psilocin per dried gram. Liberty Caps (*Psilocybe semilanceata*) have about the same amount of psilocybin but very low, sometimes undetectable, amounts of psilocin. A rule of thumb that I subscribe to is a loss of 10 to 25 percent potency after 1 year, declining to scant potency in 10 years. (*Psilocybe cyanescens* and *Psilocybe semilanceata* are notable exceptions.)

If your mushrooms are soft and pliable after you attempted to dry them, they have residual moisture and will degrade more quickly. Ideally, they should be dry within 24 to 48 hours of picking—and brittle when broken. A moisture content of 12 to 18 percent is good. Vacuum packing helps prevent remoistening and oxidation.

Preserved herbarium specimens of *Psilocybe cubensis* stored at room temperature often have no detectable psilocybin after several decades. The surprising exception, based on a few specimens, is *Psilocybe cyanescens*, which has been shown to retain its potency to about half of its typical potency up to 40 years later. *Psilocybe semilanceata* has also been shown to retain substantial potency after several decades. Bradshaw et al. speculated that the smaller initial mass of these mushrooms allows for rapid desiccation and limits endogenous enzymatic degradation.[4] Conventional wisdom was that freeze drying (also called lyophilization) was the best method of drying, although one report found that specimens of *Psilocybe cubensis* that were dried and stored at room temperature in darkness preserved more psilocybin and psilocin than deep freezing of fresh specimens.[5] Are these reports contradictory? Perhaps not, as the exact methods of dehydration may have introduced variables critical for the preservation of psilocybin mushrooms. The drying and storing method certainly affects the preservation of the active tryptamines. We still do not have a scientific consensus on the best method for drying or long-term storage.

Although the caps of *Psilocybe cubensis* have been reported to be up to twice the potency of the stems, too many variables—mushroom size, stem length, general morphology—interact to confidently predict how much degradation might occur from drying. Several published studies compared levels of psilocybin but failed to clarify whether the mushrooms analyzed were immature, sporeless fruitbodies or mature ones—or a mixture. Nor is there any detail on the drying procedure time passed from fresh collecting to analysis, light exposure, and so on. I think immature, sporeless mushrooms are more potent than mature ones from the same fruiting. More spores, less psilocybin.

At-Home Testing

Not knowing the potency of mushrooms from different species and sources, and mushrooms subjected to different storage conditions and durations, means that most journeyers depend on word-of-mouth or well-intentioned hunches. The Psilo-QTest from Miraculix, the first at-home testing kit for psilocybin content, is available at a modest cost. The test's accuracy is estimated to be within 10 to 20 percent of its true content. Nonetheless, it could be helpful for determining potency, especially for those who are sensitive to psilocybin content. When I used these test kits, the results showed just under 2 percent psilocybin in air-dried *Psilocybe cubensis* and more than 2 percent in *Psilocybe cyanescens*. However, the tests are dependent on your being able to match shades of brown between a brown liquid to a color chart, so there is plenty of room for error.

Periodically bioassaying a test batch is an empirical method for testing the potency of stored mushrooms. There is declining risk of adverse reactions with properly stored mushrooms, which means for standard dosages, over time less effect, not more, will be felt. However, for those wishing to have a predictable dosage, testing technologies, at home or by submission to analytical labs, are now becoming available. A test of this nature has enormous market potential and is critically needed for customizing dosages. An early entry servicing this need is Hyphae Labs (Patreon.com/hyphaelabs) in Oakland, California. I am not endorsing this company but appreciate their innovation and services to meet an important public need as psilocybin use increases.

Alternatively, high-performance liquid chromatography (HPLC) analysis can accurately determine potency, but this equipment costs tens of thousands of dollars, and most labs that have it won't accept psilocybin mushrooms for analysis. However, as laws are rapidly changing, with most becoming more lenient and prioritizing public safety, more testing labs are becoming available. Please periodically check for this service in your locality. Note that local laws and federal laws may not be compatible. The legal status of psilocybin possession, cultivation, use, and commerce is in a state of flux—worldwide.

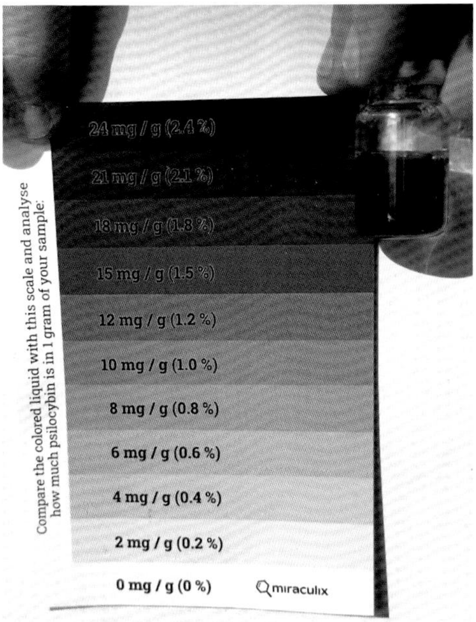

Estimating psilocybin content using a colorimetric method. Here the *Psilocybe cyanescens* mushrooms tested were stored for 4 years outdoors.

Chapter 5

Psilocybin Is Not for Everyone

Psilocybin mushrooms affect people differently. Although many of us share a similar receptivity to psilocybin, others do not. Many people find the experience of ingesting psychoactive mushrooms to be positive, profound, and spiritual. Deep reflection and connectedness ensue. Personally, psilocybin is my sacrament, the axis of my spiritual beliefs, and I use it religiously and reverently. I and many others find our psilocybin encounters to be ineffable: we have no words to fully describe the experiences. However, psilocybin experiences, although generally positive for most, can be challenging and destabilizing for some.

A modest but significant lift-off dose is felt by most people at 1 g of dry *Psilocybe cubensis* (if potency is 10 mg psilocybin in 1 g of dried mushrooms, or 1 percent), although some people feel effects when ingesting as little as 0.5 g (5 mg net psilocybin) of dried mushrooms at this concentration. A big caveat: This calculation is based on the assumption that the mushrooms have 1 percent psilocybin, but I have seen potency vary by a factor of more than four. Therapeutic macrodosing in most clinical trials is in the 25 to 50 mg psilocybin range, which is 2.5 to 5 g of dried mushrooms if the mushrooms have 1 percent psilocybin/psilocin. Note that freshly grown or collected psilocybin mushrooms may have substantially more than 1 percent psilocybin/psilocin, but on average dried *Psilocybe cubensis* has been reported to contain between 0.85 and 1.45 percent.[1] (Some analyses show nearly 5 percent psilocybin and psilocin, although infrequently.)

Microdosing is typically considered to be one-tenth of a lift-off dose, a nonpsychedelic dose, one that is not intoxicating. However, microdosing may have a noticeable effect of improving your mood, putting you more at ease, making you less confrontational or reactionary to negativity, increasing cognitive function, and perhaps improving dexterity. I think microdosing improves my sense of humor—debatably.

Two studies led by Roland Griffiths, MD, of Johns Hopkins University in 2008 and 2011 noted that while 72 percent of participants who took high doses of psilocybin in pure form found the experience to be "mystical," about one-third of the participants found that the experience

"Guardians of the Sacrament" by Mark Hansen

invoked high anxiety and fear. "Notably, 61% of volunteers considered the psilocybin experience during either or both of the 20 and 30 mg/70 kg sessions to have been the single most spiritually significant of their lives, with 83% rating it in their top five."[2]

Griffiths's team interviewed the subjects, their close friends, and family 14 months later and the ratings improved. More than 90 percent of those who had mystical experiences still described it as meaningful. The act itself of recalling the experience was, in the view of supervising psychiatrists, therapeutically beneficial. Fewer than 39 percent of the individuals who initially had negative experiences felt enduring trauma while later reflecting on the experience. It seems there was no additive, negative carryover trauma subsequently attributable to the experience itself.[3] It would be interesting to discover what a longitudinal study 4 or 10 years later might yield.

These two studies have encouraged other physicians—and the U.S. Food and Drug Administration (FDA)—to conduct and approve additional clinical studies of psilocybin to deal with diverse mental health diagnoses including addiction. However, most of these studies (at the time of writing, there are 200 psilocybin-related studies registered at clinicaltrials.gov) use pharmaceutical-grade synthetic psilocybin; only a handful use psilocybin mushrooms. Unfortunately, there has yet to be a side-by-side clinical study comparing the effects of the psilocybin molecule to a psilocybin mushroom in its natural form. Moreover, studies thus far on microdosing to date have failed to reflect real-world usage entailing microdosing several times a week.

The Mushroom Versus the Molecule

Which is "better," the mushroom or the molecule? The effects of the psilocybin molecule can often be felt more quickly because mushrooms require time to digest. Since many psilocybin mushrooms contain other tryptamines (baeocystin, norbaeocystin, norpsilocin, aeruginascin), when a person ingests a psilocybin mushroom, they also get these compounds embedded within the mushroom mass. Although these tryptamines do not cross readily through the blood-brain barrier, as psilocin does, components within mushrooms may engage with your reservoir of monoamine oxidases (enzymes that degrade amine neurotransmitters), thus reducing the enzymes' ability to partially degrade other constituents, including some tryptamines. If so, then the same calculated doses based on analyses of psilocybin in the mushroom versus psilocybin in pure form may not be equivalent in their net effects. The molecule may initially be more potent but shorter acting because of quicker absorption and ultimately less potent in terms of affecting the neuroscape as compared to the mushroom, with its longer-lasting effects and contribution of many other psychoactive components.

Psilocybin is a stable molecule that converts into psilocin in the digestive tract. When ingested, about 50 percent of the psilocin passes into the bloodstream. Following oral dosing, peak plasma psilocin levels occur 1 to 2 hours later and rapidly degrade, with a half-life of 1 to 2 hours. Psilocin is distributed throughout the body and found in various tissues, including the brain, kidneys, and liver. About 54 percent of the total metabolized psilocin is excreted as 4-HIAA (4-hydroxyindole-3-acetic acid) or psilocin glucuronide.

Comparing the effects of psilocybin mushrooms and the pure compounds psilocin/psilocybin is inherently difficult. They may not be 1:1. Subjectively and from a practical point of view, is this a difference without significance? In comparing four case reports of people taking psilocybin mushrooms versus the pure molecule, the mushroom experience was generally preferred. One participant stated that the mushroom experience was more alive and vibrant, and another said the molecule felt more medicine-y and less spiritual.[4] Which form leads to greater neuroplasticity? Which is better for depression? My bias is that psilocybin mushrooms are better than the molecule—until I see convincing evidence otherwise.

Can you add other compounds to psilocybin to enhance its effects? Adding cacao (or chocolate), for instance, a documented admixture practiced by the Aztecs, is a popular

Some caution with admixtures is advisable. Consult a physician before experimenting. For instance, combining antidepressants with psilocybin requires careful medical consultation. Some antidepressants are monoamine oxidase inhibitors (MAOIs).[5] Other antidepressants, which are more commonly prescribed, are selective serotonin reuptake inhibitors (SSRIs). A trend in the underground is to enhance the effects of psilocybin by combining psilocybin mushrooms with MAOI plants like Syrian rue (*Peganum harmala*) that contain MAOI beta-carboline alkaloids. A small percentage of psilocin is converted to 4-hydroxyindole-3-acetic acid (4-HIAA), which is implicated in treating depression,[6] by MAOs. This may be a difference without practical significance, given the other variables at play.

Thus, from a chemist's point of view, MAOIs would not substantially raise potency by enhancing psilocybin or psilocin entry through the gastrointestinal tract into the bloodstream. However, when you ingest MAOIs or some antidepressants with foods containing tyramines, the consequences can be life threatening. The concern is that some mushrooms contain significant amounts of tyramines.[7] One case report I know of has raised additional concerns. A 42-year-old man ingested 1 gram of dried psilocybin mushrooms, a combination of dextroamphetamine and amphetamine used to treat attention-deficit hyperactivity disorder (ADHD), and tranylcypromine, a MAOI prescribed for treating depression. This combination resulted in the patient soon experiencing a heart attack (myocardial infarction). He was rushed to the hospital, treated, and released the next day.[8] Combining psilocybin mushrooms with an MAOI seems like a potentially dangerous combination. Moreover, other compounds that otherwise would be detoxified in the gut by MAOIs may remain toxic; we simply do not know. Until we do, I discourage taking MAOIs to enhance the psilocybin experience or combining psilocybin mushrooms with other prescription medicines. We do not yet understand the complex interactions between psilocybin mushrooms and other drugs. Please consult a physician skilled in this subject before risking your life.

formulation used today that seems to be safe. Other admixtures, however, can be dangerous (see sidebar on page 95).

A recent prospective, longitudinal study, conducted primarily by researchers at the Johns Hopkins University School of Medicine, reported results from a survey of 2833 respondents taken 2 weeks before, 1 day before, 1 to 3 days after, 2 to 4 weeks after, and 2 to 3 months after psilocybin mushroom use. Of the total, 1182 subjects completed the 2 to 4 week post-use survey. Consuming an average of about 3 g, respondents reported statistically significant, persistent reduction in scores for depression, anxiety, burnout (fatigue and exhaustion correlated to high stress), and alcohol use and an improvement in cognitive flexibility (the ability to change thoughts about emotional stimuli).[9] Eleven percent of participants reported continuing negative symptoms. To the question: Are psilocybin mushrooms equal to or better than psilocybin the molecule? These studies suggest there are similar benefits, but more detailed comparative studies are needed as it is too simplistic to make a generalization.

Although María Sabina, the heralded Mazatec shaman, ingested some of Albert Hofmann's synthesized psilocybin and declared to Hofmann that his pills contained the "spirit of the mushroom," this reported statement needs some context. (Hofmann and his colleagues patented the process for synthesizing psilocybin and psilocin in 1959.[10,11]) María Sabina was deeply immersed in the Mazatec use of psilocybin mushrooms, a heritage passed down for centuries. The subjective experiences of those coming from an unbroken lineage of hunting, gathering, and ritualistically using psilocybin are far different from the cultural perspective and expectations of a Western scientist new to this experience. I suspect María Sabina's purported statement may have been provided apart from the benefits of her cultural context and lost in translation of this verbal exchange. I would like to know more details of this reported conversation. Nevertheless, this datapoint of one individual was an exciting affirmation that Hofmann had succeeded in his Swiss laboratory in the first-time molecular synthesis of psilocybin.

More recently, there is anecdotal evidence that psilocybin mushrooms are preferred by people who have used both the pure molecule and mushrooms of similar strength. One patient reported that, when using pure psilocybin, the onset is rapid, plateaus, and then quickly descends. On psilocybin mushrooms, the ascent is more gradual and then, after plateauing, the glide down is gentler, longer, and gives more time to reflect on the experience. Biologically, this makes sense because the mushroom must first be digested, and the longer tail end could be the entourage effect of other interacting tryptamines entering the bloodstream and reaching the brain. Our studies on neuroreceptors and neurite outgrowth (nerve regeneration) show that these associated tryptamines play important roles in neuro-activation across the receptor ecosystem. And there is likely cross-talk, meaning that receptor activation creates a cascade of effects that influence each other and other systems.

Prescription medicines require standardization, which is a challenge for all natural products. Although optimization is an understandable objective, from a practical point of view, there is a wide range of psilocybin dosages that show benefits. For instance, a range of 10 to 40 mg of psilocybin, equivalent to 1 to 4 g of dried *Psilocybe cubensis* containing 1 percent psilocybin, has proven to be beneficial for treating depression.

I have served on numerous committees to advise policy makers on the relevant issues of psilocybin decriminalization, state-level legalization for therapeutic use, cultivation methods, and possession. I have advocated for 100 g of dried psilocybin mushrooms to be legal for personal, noncommercial possession. Given the variability of psilocybin mushrooms and the fact that they degrade with time, my suggestion is that regulatory experts consider a not-to-exceed potency test with dates of analysis and expiration on the label. The worst that could happen would be a decline in potency. Less psilocybin is not more dangerous. This not-to-exceed designation offers one practical solution to the difficulty of standardization and stability. Moreover, some people are less sensitive to the same psilocybin dose than others, whether using psilocybin the molecule or the mushroom.

In a free market, where psilocybin mushrooms could be sold legally, consumers could help determine the best products. The use of mushrooms for specific medical applications requires a higher standard of use than for more general use for creativity, spirituality, or wellness. I'd like to see us strike a balance between the advantages and disadvantages of natural psilocybin mushrooms and synthetic psilocybin. My concern is that by strictly adhering to a pharmaceutical standard, we deny the majority of people the benefits of psilocybin mushrooms. Purity is an obstacle to practicality. The molecule is expensive and out of reach for most people while mushrooms are more accessible, easy to grow, and inexpensive. As Montesquieu wrote around 1726: "Le mieux est le mortel ennemi du bien" [The best is the mortal enemy of the good]. Much later, Voltaire popularized an Italian proverb worded the same way: "Il meglio é nemico del bene." In today's parlance, "Do not let the perfect be the enemy of the good." This also applies to psilocybin mushroom use. The mushroom species, its phenotype, your intention, the encompassing set and setting (one's mindset and the physical and social environment), compounded by your individuality, are all factors that can affect the outcome of any dose, no matter how carefully it is prescribed. For me, knowing that I am one in a historical continuum shared by millions over thousands of years gives me solace and confidence to maximally benefit from the psilocybin mushroom experience.

A fresh harvest of *Psilocybe cyanescens*, a potent magic mushroom, high in psilocybin and psilocin.

ALEX GREY

Chapter 6

Dosing

The three general categories for dosing are microdosing, minidosing, and macrodosing. Microdosing is nonintoxicating, whereas the subjective differences between minidosing and macrodosing are the intensities of the effects.

Microdosing

Microdosing is ingesting a small amount of psilocybin, usually one-tenth of a lift-off dose or about 0.05 to 0.1 (1/20 to 1/10) g of dried *Psilocybe cubensis*. Many people define a microdose as being "sub-sensorium," that is, a dose so small you do not sense a change in your everyday waking consciousness. I prefer the term *nonintoxicating* in this context. When I microdose, I feel better. Colors are brighter. I am happier and more pleasant to be around. I think I am funnier and more creative. I do sense a difference in my world, but I am not intoxicated. It's not like the effects of cannabis or alcohol. I think microdosing can improve neurological health with regular use, whereas one 6-hour macrodose session could be life-changing if you are aiming to come to terms with trauma, overcome addiction, and/or find spiritual equilibrium. One microdosing group, Moms on Mushrooms (momsonmushrooms.com), has an enthusiastic online following of tens of thousands of members.

The ideal microdose is specific to the individual. Although most of us won't feel any intoxication at 0.1 g of *Psilocybe cubensis*, some people will feel an effect. In studies using *Psilocybe cubensis*, potency estimates can be off by a factor of four or more unless the mushrooms used were recently analyzed. Apart from such variability, dosers show variability in their sensitivities. Any one individual's experience is influenced by a combination of factors, including genetics, diet, gut health, microbiome, food and drug intake, and surely many more yet to be determined. Each of us ought to customize our regimen. But once you know your baseline, consistent self-dosing is straightforward.

Working with the online health research platform Quantified Citizen, our research team collaborated on a prospective observational study of microdosing using a free app at microdose.me. Our first paper, published in *Scientific Reports*, surveyed people about their motivations for microdosing.[1] Surprisingly, we enrolled a nearly balanced cohort of 4653 non-microdosers and 4050 microdosers. Those who had mental health

"Flesh of the Gods" by Alex Grey

challenges were most motivated to microdose to reduce anxiety and depression. Those self-reporting no mental health challenges were more motivated to improve their cognitive performance. Of all participants, 88 percent of microdosers used between 0.10 and 0.33 g of mushrooms three to five times per week, and 90 percent used *Psilocybe cubensis*. After 30 days, a significant majority of microdosers who had mental health concerns reported dramatic reductions in depression and anxiety, whereas the non-microdosers did not. Keep in mind that the non-microdosers presumably had fewer, if any, depression and anxiety symptoms to begin with, so there were fewer symptoms to be reduced.

Although this was not a controlled study with a placebo, this collation of crowd-sourced data on microdosing to improve mood disorders is relevant. People are motivated to microdose to help improve their health. In fact, intention and expectancy are intricately interrelated. Expectancy enhances the beneficial effects of many medicines. Expectancy, the feeling of being in a community of other users and taking steps to improve your state of mind, is a factor that could enhance the medicinal benefits of microdosing. Nevertheless, in a prospective observational study of microdosing with psilocybin mushrooms,[2] participants reported significantly reduced depression and anxiety but did not show improved cognitive scores. For those seeking relief, the effects were real and microdosing worked. Clinicians can debate cause and effect, but the bottom line is the patients reported that they indeed felt better. This case showed that microdosing with psilocybin mushrooms is an effective adjunctive therapy to reduce depression and anxiety for people with those mental health issues.

Curiously, in a later study about treating moderate to severe major depressive disorder on a moderately high dose of psilocybin (25 mg) compared to the therapeutically accepted dose of escitalopram (a common medication used for depression), expectancy of taking psilocybin did not significantly affect outcome.[3,4] Hence, here expectancy was not a factor.[5] It is unclear why expectancy could influence outcome at low doses but not at high doses. Untangling the effects of intention and expectancy is complicated; nonetheless, psilocybin appears to reduce depression.

> To stay updated on the continuing publications and clinical studies on psilocybin mushrooms and other medicinal mushrooms, we update this website for physicians and researchers: mushroomreferences.com. Another free online resource is clinicaltrials.gov, which tracks registered clinical trials on psychedelics.

In general, microdosers follow a protocol of intermittent use: either 1 day of microdosing followed by 2 days off (the Fadiman Protocol), or, less commonly, 4 days of microdosing followed by 3 days off (the Stamets Protocol). Both regimens subscribe to the notion that serotonin receptors and other neuroreceptors that are sensitive to tryptamines need to regularly return to baseline, without psilocin stimulation, so they can reset and again become sensitized.

Serotonin receptors occur throughout the body, but most are located in the gut, brain stem, and prefrontal cortex of the brain, which manages perception. Psilocin and other tryptamines bind with 5HT2A and, to a lesser extent, 5HT2B serotonin receptors, which can activate a cascade of neurostimulation.[6] Psilocybin mushrooms produce both psilocybin and psilocin and some closely related tryptamines. Upon digestion, psilocybin transforms into psilocin through the

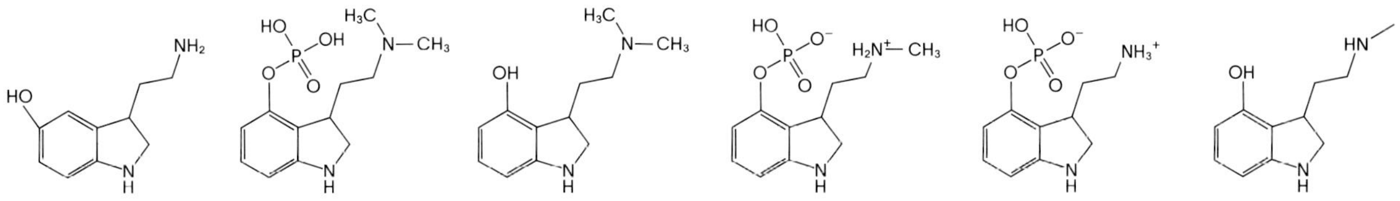

Serotonin Psilocybin Psilocin Baeocystin Norbaeocystin Norpsilocin

enzymatic removal of a phosphate group. This happens in the digestive tract so the chemical can pass through the blood-brain barrier and bind with 5HT2 serotonin receptors. Psilocin binds to ligands that allow this water-soluble compound to activate receptors inside nerve cells. This relatively small molecule crosses through the lipid-rich nerve cell membrane into the cell interior, which, in turn, stimulates nerve growth and neuroplasticity.[7]

I think microdosing can help neurological health partly because of what I have seen after spending nearly 50 years culturing mycological cell lines. Like the patterning of mycelium in a Petri dish, neurites (projections from the cell body of a neuron) grow outward in networks, cross connect, and create new surface areas on axons and dendrites embedded with highly sensitive receptors. With mycelium and tissues it is called anastomosis, and with neurons it is called synaptogenesis. Both are networks that rely on resiliency and the ability to cross communicate. My hypothesis is that, although a massive 1-day dose of psilocybin may flood serotonin (5HT2A)

Other closely related but less understood tryptamines that stimulate neuroreceptors include baeocystin, norbaeocystin, norpsilocin, and aeruginascin.[8] Until recently, neuroscientists did not agree that new brain cells, called *newborn neurons*—not to be confused with *postnatal nerve cells*—could be created. The presumed absence of newborn neurons after early childhood has long been dogma in neuroscience. But now we know the hippocampus can make new neurons and that stem cells can differentiate into neurons. Evidence is still limited, yet neurogenesis is now documented even in people with dementia or early-stage Alzheimer's disease.[9] What excites me is that psilocybin and neurochemically related tryptamines can help create newborn neurons from stem cells. We have evidence that they stimulate nerve growth factors, brain-derived neurotropic factors that not only help create new neurons but also extend neurite outgrowth and networking, rejuvenating existing neurons. Even if these psilocin- and psilocybin-related compounds are proven to stimulate neurogenesis to a lesser degree than stimulating the cross networking of neurons (neuroplasticity), the combination of beneficial effects in the neuroscape could be a game changer in treating an array of neuropathies. We are still in the earliest stages of this branch of neuroscience.

receptors, the neurological benefit is not the same as titrating the same dose over a longer period. In fact, it has become increasingly popular to do both: take a macrodose one day and then follow up with a regimen of microdosing. Time will tell as the science progresses.

Once most people have had one macrodose of psilocybin, they do not want to repeat an experience of such intensity for months, sometimes years. However, nonintoxicating microdosing is attractive. The theory goes that microdosing helps to reinforce the new neurological pathways established during the initial macrodose. These new pathways are groomed to become part of the new resident neural operating network. The best, strongest indicators will come not from healthy normals, but initially from those who suffer from difficult-to-treat mood disorders.

My research team has focused on this subject for several years. The half-life of psilocin is no more than 1.8 hours, meaning that in 24 hours there is only 0.000097 of the original dose, or less than 1/10,000th, still present in your bloodstream. After a day or two, dosers have returned to baseline in their plasma. This metabolic reset does not account for myriad events that psilocin can precipitate inside a nerve cell, including receptor occupancy and receptor internalization, which leads to tolerance and reduces the efficacy of multiple high doses. But it could explain why, when I once consumed high doses of psilocybin mushrooms 4 days in a row, I barely felt any effect by day 4. In contrast to macrodosing, with microdosing fewer receptors are awashed.

One study showed that approximately 2 mg of psilocin only had a 20 percent occupancy rate in 5HTA receptors, whereas up to 30 mg of psilocin had a 50 percent occupancy.[10] This means that latency periods may be reduced with microdosing compared to macrodosing, and maybe with less occupancy there is greater sensitivity because the receptors are not swamped as they are with higher psilocybin doses. By titrating psilocybin, psilocin, and other tryptamines over months, microdosing can provide a repeated low-level stimulus for neurogeneration. Neurites take time to grow, and modulating the neurogenic properties of psilocybin tryptamines may show greater benefit for nerve cell extension than one big dose.

In 2022, we published our second paper in *Scientific Reports* on the observed effects of microdosing.[11] Microdosers reported reductions in anxiety and depression compared to non-microdosers. One other number jumped out: participants aged 55 and older who combined psilocybin mushrooms, vitamin B_3 (niacin), and Lion's Mane mushroom mycelium (a formula I call the Stamets Stack) showed an increase in psychomotor skills as measured by the tap test used by physicians to measure recovery from traumatic brain injuries and the neurodegenerative consequences of Alzheimer's disease, Parkinson's disease, dementia, and other neuropathies. There are various types of tap tests, but the one we used measures how many times a person can alternatively tap their two forefingers on a surface with one hand in 10 seconds. The chart opposite summarizes what the data showed after 30 days of tap tests in participants who used the Stamets Stack, microdosed psilocybin alone, or did not microdose.

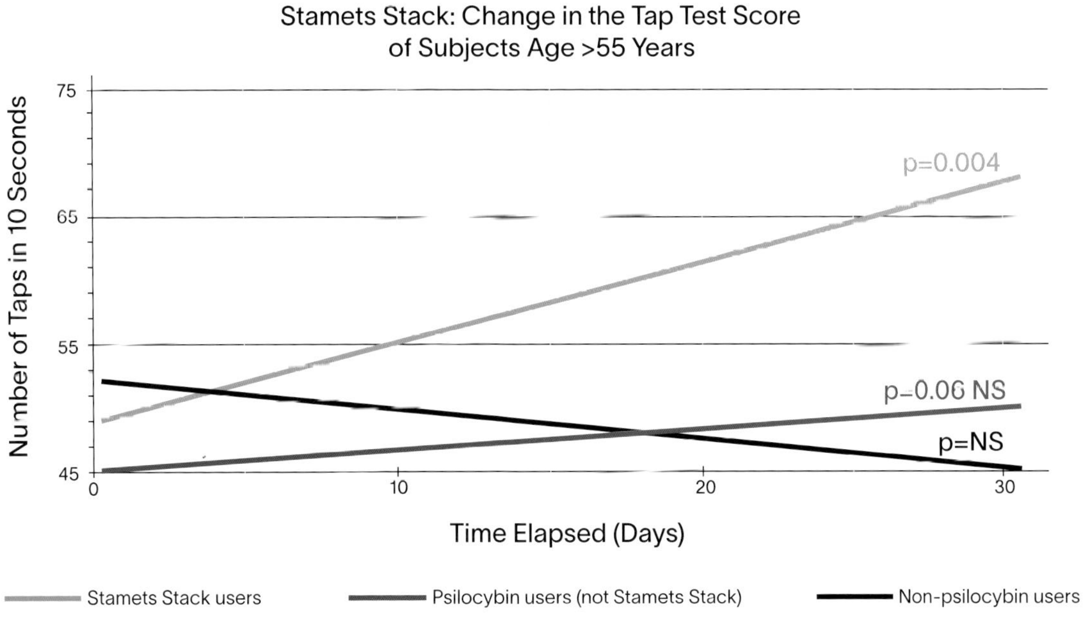

Participants aged 55 and older who used the Stamets Stack increased their tapping agility from about 48 taps to 68 taps in 10 seconds over 30 days ($p = 0.004$, meaning there is a 1 in 250 chance this signal is random). Note that non-microdosers and microdosers using psilocybin mushrooms by themselves or with any other admixture in any other form had no significant improvement in tapping performance. This improvement in psychomotor frequency suggests an effect that cannot be explained by expectancy. To ideate tapping to physically performing the tap test involves at least seven regions of the brain and neuromuscular system, suggesting that the Stamets Stack has physio-neurological benefits independent of the subject's expectancy. One caveat: We do not know the effect of niacin (B_3) and Lion's Mane independently due to study size. More research is needed.

The Stamets Stack

The Stamets Stack is a formulation I came up with to help maximize the neurological benefits of psilocybin when added to other compounds, in this case Lion's Mane (*Hericium erianceus*) mushroom mycelium and vitamin B_3 (niacin or nicotinic acid). An exciting aspect of ingesting psilocybin mushrooms is their potential for neuroregeneration (regrowing existing neurons), increasing neuroplasticity (increasing interconnections between neurons), and neurogenesis (the birthing of new neurons). Adding Lion's Mane mycelium and niacin appears to enhance this process. Our research shows synergism of these components that not only extends neurite outgrowth in vitro but also increases synaptic crossings (synaptogenesis) and neuroplasticity by stimulating nerve growth factors to bind with tropomyosin receptor kinase B (TrkB) receptors, which induce survival and differentiation of cells, including neurons. Psilocin was found to stimulate the binding of endogenous brain-derived neurotrophic factors in the synapses between neurons. Stimulation of TrkB receptors is considered one of the primary mechanisms of antidepression drugs.[12]

In one study, surprisingly, psilocin was reported to bind with TrkB receptors more than 1000× more than many drugs at a therapeutically high dose. If this 1000× binding occurs after high doses of psilocybin, microdoses and minidosing may show stimulation of brain-derived neurotrophic factors binding to TrkB receptors greater than most FDA-approved antidepressant medicines. Psilocybin is poised to become a new antidepressant, neuroregenerative medicine that could rival SSRI antidepressants such as escitalopram (Lexapro), fluoxetine (Prozac), paroxetine (Paxil, Pexeva), and sertraline (Zoloft).[13]

I have met several people who have weaned themselves off antidepressants by microdosing with psilocybin and then, in some cases, ceased microdosing regularly, only to begin again on an "as needed" basis. Many were evangelical in their enthusiasm after finding a new lease on life. They overcame anhedonia and could feel pleasure again. Controlled clinical trials are needed to get a better sense of this effectiveness.

I decided to combine psilocybin mushrooms, vitamin B_3 (niacin), and Lion's Mane mycelium because I thought by adding niacin, a vasodilator, it might help offset the vasoconstriction of psilocybin and ensure that more psilocybin was delivered via the blood to the nervous system. Niacin is also known to be neuroregenerative, passing through the blood-brain barrier. Studies have shown that when Alzheimer's disease and dementia patients take niacin it can help to slow the progression of these types of neurodegenerative diseases.[14,15] Since niacin also aids the conversion of the amino acid tryptophan into serotonin, the pairing of niacin and psilocybin, I thought, could enhance psilocybin and psilocin's benefits synergistically. I think niacin is a potentiator not only for psilocybin, but also likely other tryptamines, including baeocystin, norbaeocystin, norpsilocin, aeruginascin, and dimethyltryptamines like N,N-dimethyltryptamine (DMT) and 5-methoxy-N,N-dimethyltryptamine (5-MeO-DMT). Niacin might help to enhance the benefits of MDMA, LSD, ketamine, and other psychedelics. Notably this is a speculative hypothesis, yet to be proven.

Lion's Mane (*Hericium erinaceus*) is an edible and medicinal mushroom used by humans for hundreds of years. Commonly found on aging hardwood trees in the United States but increasingly rare, for instance, in Great Britain, its mycelium and mushroom fruitbodies contain complex potent compounds in a constellation around erinacines and hericenones. As first reported by the Japanese researcher Hirokazu Kawagishi, Lion's Mane mycelium's erinacines is a presumed defensive mechanism from contact with bacteria. Kawagishi extrapolated that erinacines would help nerve health and discovered that they and related compounds, in fact, do stimulate nerve growth factors[16,17] and brain-derived nerve growth factors, as confirmed by many other researchers. Lion's Mane mycelium extracts also induce the recovery of myelin, the conductive sheath covering the axons of nerves that facilitates neurotransmission. Several, albeit small, clinical studies show benefits of Lion's Mane ingestion by people with mild cognitive decline and depression.[18,19] Preliminary evidence in nonhuman models also suggest a possible benefit for treating Parkinson's disease.[20]

By combining these components, I intuited an effect of benefits augmenting the activity of each ingredient beyond its individual or additive effects. This is known as synergism. The following baseline formula for the Stamets Stack could continue to be built upon, as there are many other potentiating components. I like to take this stack about one hour before a mid-afternoon nap or in the evening before going to bed. I have noticed not only a heightened state of active dreaming, but I also feel more rested and renewed on awakening. Others have reported the same. I sense this practice is neurologically beneficial.

Consuming carefully formulated mushroom chocolates is a convenient and popular way of microdosing. Keep them out of the reach of children and unaware adults, and carefully label with "Do Not Eat!"

The Stamets Stack Baseline Formula

Ingredient	Amount
Dried psilocybin mushrooms with 1 percent psilocybin/psilocin	50–100 mg (1/20 to 1/10 g)
Lion's Mane mushroom mycelium	100–500 mg (1/10 to 1/2 g)
Vitamin B_3 (niacin, nicotinic acid)	25–50 mg (1/40 to 1/20 g)

My original formula called for up to 200 mg of niacin, however, the flushing reaction (tingling, excitement of nerve endings, skin reddening due to vasodilation) was uncomfortable for some. For many, 25 mg is much more tolerable. Also, up to one-third of a gram (333 mg) of dried psilocybin mushrooms can be used. I am super sensitive, so one-twentieth is my preference. Many people have told me that, after repeatedly microdosing with psilocybin and niacin, the noticeable effects of niacin diminish or are barely noticeable. Some people don't feel any change in perception. People's reactions vary.

Delivery Systems for Microdosing

I can't stand the taste of many psilocybin mushrooms. As soon as I taste them, I get queasy. I have a friend, however, who loves their flavor. We are all different—not only in our flavor preferences, but also in our biochemistries as we merge with the psilocybin mushroom's biome.

Thanks to the Aztecs, we know that psilocybin mushrooms and chocolate pair well. The mushrooms are much more palatable when a measured amount is blended with melted chocolate to make a single dose. Pouring melted chocolate into wells of candy forms (see photos above) that contain a measured amount of psilocybin mushrooms and/or other ingredients like Lion's Mane mycelium and/or niacin is a convenient way to make batches for daily microdosing. Note that weighing and premixing all together before pouring is easier.

Another excellent way to ingest pulverized mushrooms is in tea with honey. Lemon, lime, turmeric, mint, and spice teas mask the unsavory flavor of psilocybin mushrooms quite well. Once you find a flavor and dosage combination that works for you, stick with it. Culinary adventures

A warning to those who are eager to experiment: People advertising themselves as shamans and healers probably are not. True shamans are humble and do not seek customers. The burgeoning interest and commercialization of psychedelics inspires entrepreneurs and business plans that can be at odds with responsible mushroom messaging. Successful models, on the other hand, include selected legal psychedelic retreat centers in Jamaica, Costa Rica, and the Netherlands that have honed their skills in creating comfortable destinations that are visually and musically enriched. Unfortunately, with the resurgent popularity of psilocybin, consumers should be wary. This new psilocybin-rich mind field could become a minefield of undesirable consequences. There is one overarching solution, however: Psilocybin mushrooms should be legalized for personal use. Psilocybin access should be a fundamental civil right: the freedom of consciousness. We need a peace treaty with psilocybin.

with psilocybin mushrooms run risks as previously noted with MAOIs. Please make sure your admixtures do not have MAOI activity.

It's important to keep psilocybin mushrooms out of the reach of children or unaware adults. Labeling your chocolates as "laxatives" can dissuade would-be nibblers, as can storing them in a safe and secure location. Accidental ingestion of psilocybin chocolates is an all-too-common story. Please be responsible.

The Museum Dose: Mini or Medium Dosing

Minidosing, also known as medium dosing, means taking a low, slightly intoxicating dose of psilocybin. From my informal observations over the past few years, a minidose is 0.3 to 1 g of dried *Psilocybe cubensis* or an equivalent that has 1 percent psilocybin, equivalent to 3 to 10 mg psilocybin. This is also called a museum, or lift-off, dose—so called because trekking through museums, with their rich, colorful, diverse art on a minidose, is popular in some circles.

Museums are mellow places where people tend to be respectfully distanced and quiet. Museum dosing is often done with friends and makes for a great deep dive into the art, archaeology, paleontology, history, and cultural diversity. Museum trippers are easy to spot. They may wear sunglasses indoors and seem unusually happy. When taking psilocybin, the pupils of our eyes sometimes dilate—our visual field is brighter—so significantly that the whites (sclera) of our eyes disappear. Suppressing laughter with friends is often a challenge too, especially with newbies.

For the more athletically inspired, a museum dose can be a "sports dose" as its proponents espouse improved psychomotor benefits: better coordination, timing, teamwork, and performance. These low doses are much more manageable than high doses, which require careful planning and preparation. While minidosing, I like to go mountain biking while listening to great music. I find that I am more coordinated. I fall into a Zen-like flow state—hyper aware of the rapidly streaming visual field that passes by me—but I feel unusually dexterous. (Note to self: Be careful.)

I also know minidosers who have accomplished remarkable feats of hand-eye coordination that strongly hint of increased psychomotor skill performance—in baseball, basketball, soccer, tennis, archery, music, and martial arts, to name a few. But I also know people who have had accidents as they pushed minidosing beyond their safety limits. Reading contracts, signing legal documents, or engaging with unaware or unsupportive non-dosers is not advised. Best to stick together with experienced and responsible dosers and watch out for one another.

Minidosing to enhance performance is controversial for more conservative minds, but the fact is many people embrace medium dosing to enhance performance and coordinate teams of athletes and others involved in team activities. Medium dosing is also practiced by musical groups, who claim it allows them to perform better and to effortlessly sync with the others in the band. The Dead and formerly, The Grateful Dead, are two of my favorite examples.

For me, the low end of medium dosing is about 3 mg of psilocybin and the high end is around 6 mg, or about one-third to two-thirds of a gram of dried mushrooms. From my personal experience, 8 mg of psilocybin (i.e., 0.8 g of *Psilocybe cubensis* containing 1 percent psilocybin) is pushing the outer limits of a "medium" dose—it's strong enough to help me focus internally, but not a full-blown spiritual, cosmic adventure into oceanic boundlessness. Even at this dosing, I want to be in a safe place, with no obligations as a caregiver to others, and certainly not to drive, use chainsaws, or engage in any activity that could endanger myself or others.

Keep in mind that a medium dose quickly becomes subperceptual when used on consecutive days. A medium dose on Monday could be a microdose by Wednesday. Each person may react differently.

Macrodosing

Macrodosing can launch you into the depths of your soul and may challenge and change your beliefs and your views of what is real and true and what matters. After macrodosing, people typically feel more content, cooperative, wise, humble, and at peace. Big things that seemed important before your macrodosing voyage are now less pressing and no longer define your life.

Medical macrodosing, also called psilocybin-assisted therapy, seems effective at prompting an epiphany that helps patients understand that they no longer need to consume addictive drugs or behave antisocially and self-destructively. Expanding on the pool of anecdotal claims over decades, recent clinical studies support that psilocybin-assisted therapy can help addicts stop cigarette smoking, binge drinking, or opioid use—in some cases, within a day or two.[21]

Some people wonder whether psilocybin is addictive. In fact, psilocybin has been described by experts as the "anti-addictive drug" and showed the least antisocial harm of twenty commonly used drugs in the United Kingdom (see chart on page 108).

Macrodoses usually begin at the equivalent of 2 to 3 g of *Psilocybe cubensis* at 1 percent psilocybin/psilocin and above, which is equivalent to or greater than 20 to 30 mg of pure psilocybin/psilocin. However, dried *Psilocybe cubensis* and *Psilocybe cyanescens* can have nearly 4 percent psilocybin, in which case 2 or 3 g of the mushrooms would contain 80 to 120 mg of pure psilocybin/psilocin. This is approximately 2.5 to 5 times higher than a therapeutic macrodose used in many clinical studies using pure psilocybin. Any dose equivalent to more than 25 mg is considered by most experienced users to be embarking on the "hero's journey." Macrodoses can be as high as 25 g of mushrooms, but typically

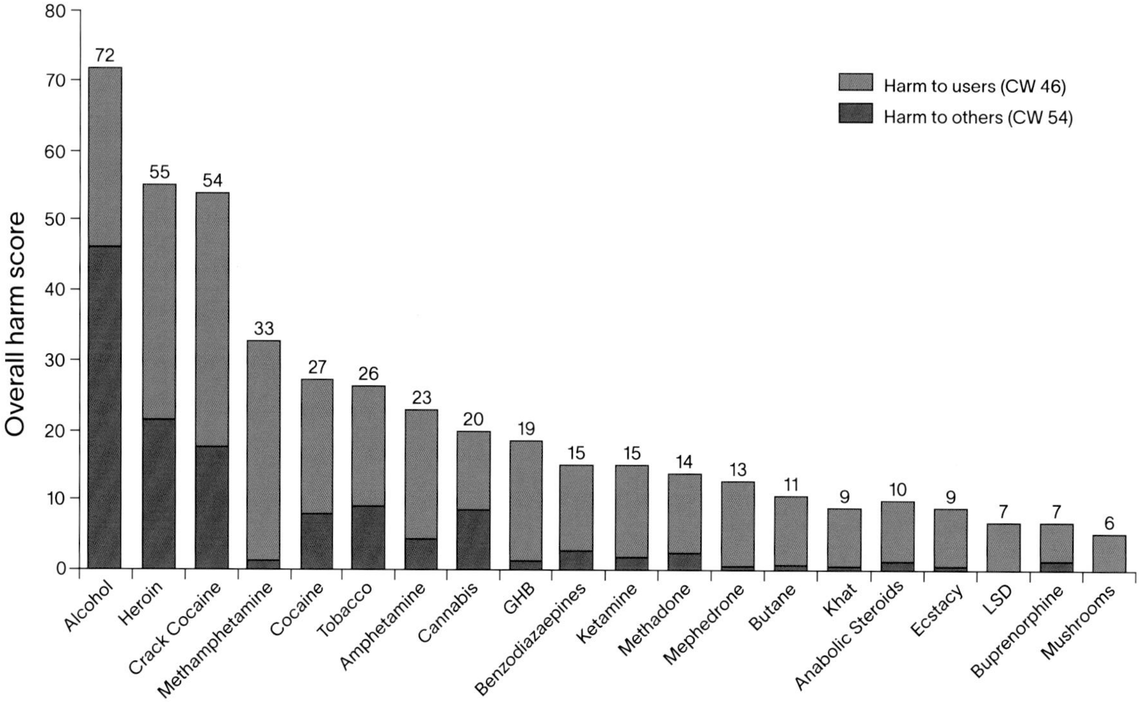

Comparison of commonly used drug and their harm potential to users and others. Image Credit: Nutt, D. J., King, L. A., Phillips, L. D. 2010. Drug harms in the UK: a multicriteria decision analysis. *The Lancet* 376(9752):1558–1565.

anything more than 4 g is characterized by the journeyer needing to lay down; it requires one's total attention to navigate and take in the stimuli cascading as the floodgates of the senses are opened. I experience a state of oceanic boundlessness at 4 g, equivalent to 40 mg of psilocybin at 1 percent. Of course, each person responds individualistically. Some people are more sensitive than others. The experience typically begins within 10 to 20 minutes and can last between 4 to 6 hours. That's a long time to reflect on one's life. Although macrodoses of psilocybin adversely affect coordination in most people, others can become mind-bogglingly coordinated.

If you are inexperienced in taking psychedelic mushrooms, it really helps to have a seasoned compassionate therapeutic guide (a health care professional, skilled guide, or a trusted friend) with you throughout your journey. This is especially essential should you have a difficult time dealing with past trauma that resurfaces. I do not recommend that novices try journeying doses of psilocybin by themselves. Being in a safe setting and preparing emotionally is key for maximizing the good and minimizing the bad during a psilocybin macrodose. A psilocybin journey is serious and could be one of the most significant milestones of your life. Before you begin, I suggest you read a guidebook such as *Manual for Psychedelic Guides* by Mark Haden, *Conscious Medicine: Indigenous Wisdom, Entheogens, and Expanded States of Consciousness for Healing and Growth* by Françoise Bourzat and *Have A Good Trip: Exploring The Magic Mushroom Experience* by Eugenia Bone.

Fasting before macrodosing journeying reduces risks of adverse physical events. Eating before dosing can influence the quality and intensity of the experience. Indigenous peoples and modern-day psychonauts abide by a common rule of safety: low food intake on the day of journeying—or no food intake for at least 4 hours before. An increasingly common practice is to precondition the gut microbiome by ingesting Turkey Tail (*Trametes versicolor*) mycelium-based supplements for several days beforehand. One double-blinded, placebo-controlled study showed that Turkey Tail mycelium-based supplements enhanced beneficial bacteria while displacing inflammatory bacteria.[22] The gut microbiome–mind connection is a subject of increasing interest to help explain why some people are super sensitive to psilocybin mushrooms, whereas others are not.

Macrodosing, Medium Dosing, and Microdosing: A Hybrid Regimen

An increasingly popular dosing strategy is to take an infrequent macrodose (2.5 to 5+ g of dried *Psilocybe cubensis* at 1 percent psilocybin) followed by frequent microdoses (50 to 100 mg of same). By infrequent, I mean macrodosing one or two times a year, and then microdosing several days per week for months, or until not needed.

Another hybrid regimen is to take a macrodose, and then take a "refresher" medium dose every few months that is one-quarter to one-half of the original macrodose.

The macrodose is a revelatory, therapeutic, intense experience. Often, macrodoses lead to a breakthrough in one's patterns of thinking and behavior accompanied, presumably, by the creation of new neural pathways. Gül Dölen's lab found that there is a critical period after a high dose. By following a high dose of psilocybin

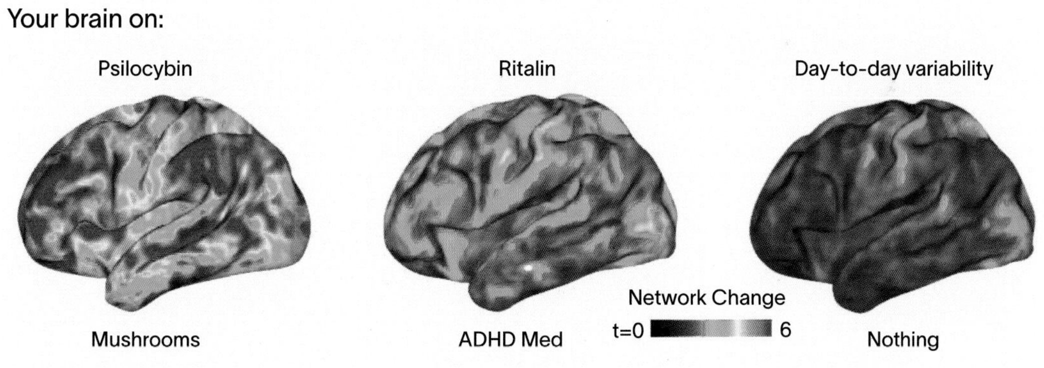

Psilocybin massively disrupted functional connectivity (FC) in the cortex and subcortex, acutely causing more than threefold greater change than methylphenidate (Siegel et al. 2024).[23] Methylphenidate, which is commonly sold as Ritalin, is the positive control. This study showed that psilocybin de-synchronizes neural networks, enabling pro-plasticity, the creation of new network signaling between synapses. This is neuroanatomical evidence that psilocybin helps to restructure the brain. Psilocybin can change your mind.

Many of us know people who, after a heroic dose of psilocybin, became a new person. Gül Dölen and her lab in the Department of Neuroscience at Johns Hopkins University School of Medicine demonstrated that during a psilocybin journey, a critical period (a window of time when the brain is especially sensitive and malleable) reopens, allowing for behavior modification.[24] This effect is shared across multiple psychedelics, including psilocybin, although in some cases the critical period reopening is dependent on 5HT2A binding (psilocybin, LSD), and in some cases it is independent of 5HT2A binding (MDMA, ketamine, ibogaine). They demonstrated that psilocybin and other psychedelics restore the ability of synapses to learn (metaplasticity) but did not induce the kind of pathological learning associated with addiction and cancer (hyperplasticity). Additionally, during the psychedelic-induced critical period reopening, the extracellular matrix (the proteins and sugars that form the glue between synapses) is reorganized.

When patients were given macrodoses of pure psilocybin followed by brain scans with MRI (magnetic resonance imaging), researchers at Washington University School of Medicine discovered something not seen with other psychedelics: psilocybin reduced connectivity between the anterior hippocampus in the cortex and the default mode network, which is central to our sense of space, time, and self.[25] Psilocybin caused desynchronization between the cortex and subcortex. This desynchronization allowed an opportunity for neuroplasticity, forming new synaptic connections. In essence, the result was neuroanatomical. Psilocybin physically alters the connections between neural networks, providing patients with opportunities to rewire their brains, break out of habitual behavior, and emerge as a new person. Truly, as Michael Pollan noted in his book *How to Change Your Mind*, psilocybin can help you change your mind.[26]

with microdoses, those new neural pathways are revisited, making them more dominant (or resident) than the previously rooted mainstream pathways. Reflecting on your macrodosing experience could be one way to animate those new pathways in your head.

How combinations of microdosing, minidosing, and macrodosing with psilocybin and psilocybin mushroom species—and their periodicities—improve neurological functioning is yet to be clinically determined.

A seminal study by Roland Griffiths's team at Johns Hopkins University showed that patients who were interviewed more than a year after ingesting a high dose of psilocybin benefitted therapeutically merely by remembering that journey. Even beyond the neurogenic stimulation of brain-derived neurotrophic factors and nerve growth factors, restaging the macrodose by microdosing is a portal of willful return that repaves newly built neural pathways.

This regimen shows benefits related to addiction, including use disorders of alcohol, opioids, and tobacco. Macrodosing with psilocybin as a means of treating depression and other mood disorders, supported by follow-up therapy, is also well supported by several clinical trials.

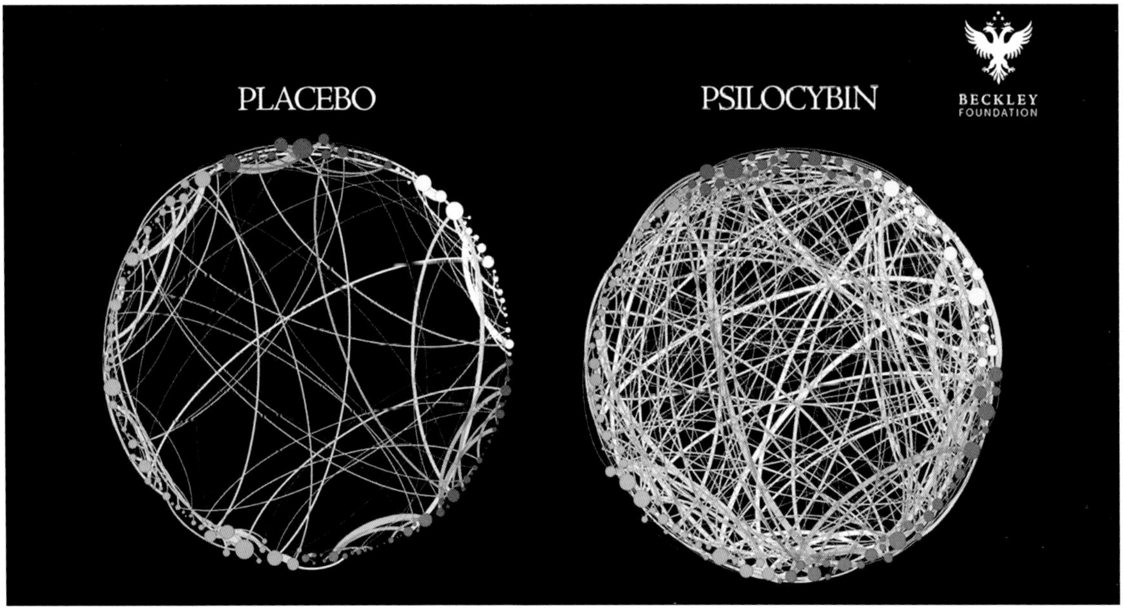

The brain on psilocybin. A simplified visualization of connectivity between functional areas and networks in the brain during the resting state under placebo (left) and psilocybin (right).[27]

Notable Clinical Studies on Psilocybin Use

Psilocybin is a versatile medicine. What other medicine could help alleviate depression, anxiety, and addiction (tobacco, alcohol, opiates) and reduce violence, crime, and inflammatory diseases while stimulating regeneration of nerves? Psilocybin fundamentally improves our neurological health. Ultimately, I think psilocybin makes nicer, healthier people.

Many people prefer to engage with the mushroom rather than just the psilocybin molecule. Ingesting psilocybin mushrooms connects you to a lineage of deep history of human use, a sense of community that you share with others on this journey. But I know other people who prefer the feelings of safety and security that come with ingesting a pharmaceutically pure psilocybin molecule within the confines of a medical facility, staffed with trained professionals, alleviating their fears should they have a challenging experience.

Since psilocybin could be beneficial across a wide spectrum of consumers, I think there is space for both approaches. The difference is that psilocybin mushrooms are gifts from nature. Synthetic psilocybin is a pharmacological copy of just one molecule in the organic matrix of psilocybin mushrooms. Are they essentially the same? My opinion is that there are significant differences that influence availability, affordability, and outcome. Psilocybin mushrooms can be grown for about a dollar per fresh pound, and only one-tenth of a fresh pound is needed for a high therapeutic dose, while pharmacological psilocybin currently remains far out of reach pricewise, costing thousands of dollars for an equivalent dose.

As of this writing, there are 200 clinical trials registered at clinicaltrials.gov using psilocybin as

"So I'm guessing we're in the placebo group."

I love this cartoon by Paul Noth underscoring the futility of having a placebo in a high-dose psilocybin clinical trial.

an investigational new drug (IND), as mandated by the FDA via the National Institutes of Health. No doubt many more will be added as the therapeutic potential of psilocybin is further explored. Investigations and studies are likely to refine and expand the potential benefits of psilocybin. Given the selection criteria needed for enlisting patients into clinical trials, many patients who could benefit may be excluded if they have disqualifying preconditions. Clinical trial protocols often require a subset of patients who are unlike the general population. The therapeutic use of psilocybin and psilocybin mushrooms will be expanded and refined as personalized medicines, customized to the individuals in need by skilled health practitioners.

Double-blinded, placebo-controlled clinical trials are deemed the gold standard for testing the effects of new medicines. The premise is that patients should not be able to tell the difference between the medicine and the placebo. However, for mental health studies, choosing a medicine and a placebo that cannot be easily detected by patients is challenging, particularly when macrodosing with psychedelics like psilocybin. In 20 minutes, most people feel an increasing wave of intoxication from psilocybin, and 1 hour after ingestion the experience is highly noticeable.

Another confounder with clinical studies is the hospital-like environment. Some patients and test subjects would find this setting itself to increase anxiety. Patients entering a doctor's office experience higher than normal blood pressure levels. Taking the same dosages of psilocybin within the comfort of one's own home, with accustomed surroundings and safety and privacy, and with skilled friends or guides may augment positive outcomes versus dosing in an institutional environment.

The influence of the environment has always been a strong factor in clinical studies. When dealing with any mood or mental health disorder, the panoply of variables can skew findings. We need highly skilled therapists who are themselves psilocybin-experienced to best guide patients in the studies. Psilocybin-naive physicians are at a disadvantage in understanding the experience their patients must navigate. Physicians' sympathy or apathy may also influence outcomes. On the other hand, psilocybin-experienced physicians are not only more informed, but can be more empathetic and trusted. When the late, great Roland Griffiths, MD, from Johns Hopkins University, who conducted many controlled clinical trials with psilocybin in carefully controlled hospital settings, was asked the ideal setting for using psilocybin personally, he said "outside, under stars with a group of three or four friends."[28]

Prospective observational studies offer useful insights that can lead to clinical trials. If a person stops smoking, drinking, or using opiates after one or two sessions of ingesting high doses of psilocybin, do we really need a placebo to show that psilocybin can reduce addiction?

Several of the completed and ongoing clinical studies involving psilocybin were first informed by retrospective observational studies.

Physicians often use best-case outcomes and meta analyses to explore whether there is a cause and effect for a new medicine to treat a disease. As more people report to mobile apps like microdose.me, I hope more data will emerge to help test whether these indications are more than a coincidental association, but actually cause and effect. Here are some of the psilocybin studies at clinicaltrials.gov that I find particularly interesting and that show indications worthy of further research.

TOBACCO CESSATION

One open label study conducted at Johns Hopkins University enrolled a small cohort of 15 smokers who were given two doses (20 and 30 mg) of psilocybin, followed by talk therapy, which resulted in 10 (67 percent) of the participants being smoking abstinent 12 months later.[29] Nicotine is one of the most addictive drugs, harming individual smokers and taxing the health care system. Psilocybin may be an antidote for cigarette addiction. Currently there are several clinical trials exploring how psilocybin treatment can help alleviate tobacco addiction listed at clinicaltrials.gov.

ALCOHOL USE DISORDER

In a study designed to test the effectiveness of psilocybin for treating alcohol use disorder, 93 subjects were enrolled in a randomized clinical trial comparing two psilocybin treatments (25 to 40 mg) versus two diphenhydramine treatments, the placebo control. Both groups also received therapy. Thirty-two weeks later, the level of heavy alcohol consumption was significantly reduced in the psilocybin group. The percentage of heavy drinking days during the double-blind period was 9.7 percent for the psilocybin group and 23.6 percent for the placebo (diphenhydramine) group.[30] Currently there are sixteen studies on clinicaltrials.gov exploring how psilocybin therapy can treat alcohol use disorder.

DEPRESSION AND ANXIETY

In 2019, the FDA designated psilocybin as a breakthrough therapy for treating severe depression.[31] The breakthrough therapy designation is a "process designed to expedite the development and review of drugs that are intended to treat a serious condition and preliminary clinical evidence indicates that the drug may demonstrate substantial improvement over available therapy." There are seventy-six notable studies at the time of this writing for treating depression and anxiety with psilocybin at clinicaltrials.gov.

Roland Griffiths et al. published a randomized, double-blinded, placebo-controlled study of 51 cancer patients with anxiety and depression who showed enduring reduction of their symptoms 6 months after a two-dose treatment of 22 and 30 mg of psilocybin.[32] In a joint research project, Sandeep Nayak, Brandon Weiss, and their colleagues enrolled a total of 59 patients, 30 of whom were assigned to a group that received psilocybin treatments (25 mg) and 29 who took escitalopram for 6 weeks (escitalopram is a prevalent SSRI drug approved by the FDA for treating depression). According to several depression scales, the psilocybin treatment group showed improvement in overcoming much of their depression-related symptoms with results that were "certainly not inferior" to the escitalopram group.[33] Weiss et al. also noted that "post-escitalopram changes in personality were significantly moderated by pre-trial positive expectancy for escitalopram, whereas expectancy did not moderate response to PT [psilocybin therapy]."[34] The role of expectancy influencing psilocybin-assisted therapy has been hotly debated, especially as it relates to microdosing. This study suggests that expectancy had negligible influence on outcomes. If expectancy was not a factor with a macrodose, why would expectancy influence the outcomes of a microdose regimen? Even if expectancy was a factor with microdosing, it could be a collateral benefit. I wonder if these differences are artifacts of our seeking a narrow interpretation of causes and effects. And yet all doctors count on expectancy to enhance the outcomes of medicines with their patients. Doctors who care about their patients uphold best practices in psychedelic medicine.

Natalie Gukasyan et al. conducted a clinical study for major depressive disorder (the form of depression that many suffer from) in which 24 patients finished the psilocybin arm after 1 year, with 50 percent showing significant durable reduction in depression compared to the placebo group.[35]

But psilocybin is not a panacea for treating all types of depression. Some patients do not respond as well as others. For the more severely challenged patients who suffer from treatment-resistant depression, Goodwin et al. enrolled in a Phase II study a total of 233 participants assigned to a 25 mg psilocybin group, 75 to a 10 mg group, and 79 to a 1 mg group (control). They reported that "adverse events occurred in 179 of 233 participants (77%) [in the 25-mg group] and included headache, nausea, and dizziness. Suicidal ideation or behavior or self-injury occurred in all dose groups." Reduction of depression was comparatively modest, although this patient group was more severely depressed and difficult to treat compared to those in other studies.[36] These findings underscore that some patients who have treatment-resistant depression may not benefit as much as those with the less severe diagnosis of major depressive disorder when treated with synthesized psilocybin. Generally, however, psilocybin appears to help reduce depression.

For people severely challenged with depression, additional therapeutic support is essential for maximizing positive outcomes, customized to their needs. After a macrodose session, talk therapy with a clinically trained professional is essential to optimize the benefits and minimize harm for patients. Those who are not chronically depressed appear to generally benefit from psilocybin, and reflecting on experiences with friends and family helps consolidate the positive benefits of psilocybin.

Retrospective Observational Studies on Psilocybin

Retrospective observational studies are not randomized or interventional, but rather observe trends in a population. The most common types are surveys. For instance, some observational studies take place with patients in hospitals or with prisoners. These are not controlled studies but are rather designed to glean "signals from the noise."

Following are five studies that I find fascinating and that may lead to more controlled clinical trials. Some, by their very nature, would be quite expensive to conduct as they would have to track behavior over many years. Undoubtedly, there will be more such studies to come.

Nature Relatedness: A prevailing theme of psilocybin experience is an increase in what's called "nature relatedness."[37,38] Most psilocybin psychonauts become more sensitive to human impacts on ecosystems, the importance of becoming better stewards of biodiversity, and the urgency of protecting the environments that sustain us. In so doing, we ally with psilocybin mushrooms and aid their survival by growing and using them.

Matthais Fortsmann et al. surveyed 3817 people about their use of various psychedelics and attitudes about environmental issues to assess the correlation between psychedelics use and nature relatedness.[39] Comparing psilocybin, LSD, mescaline, *Salvia divinorum*, ketamine, and ibogaine, they discovered the only clear reliable predictor of increased nature relatedness was the use of psilocybin mushrooms. This is consistent with experiences of many psilocybin users I know: We become more committed to protecting the ecosystems that provide for us and promote biodiversity. What is unique about psilocybin here? Is it just the molecule? Or the molecule and the mushrooms that connect us to the long history of use that makes this psychedelic stand out? When I do a high dose of psilocybin mushrooms, I sense that I am aligning with ancestral knowledge keepers, entering into a unified field of shared consciousness. I no longer feel alone. I feel one with Nature and the Universe.

Crime: Based on the responses of more than 480,000 U.S. adults to the National Survey on Drug Use and Health (2002–2014), psychedelic use, particularly weighted by the positive influence of psilocybin, was associated with a population-wide reduction in criminal behavior, with 12 percent decreased odds of assault in the previous year, 22 percent decreased odds of arrest for a property crime, 27 percent decreased odds of larceny/theft, and 18 percent decreased odds of arrest for a violent crime.[40]

Domestic Violence: In a survey of 1266 community members aged 16 to 70, men who had past experiences with either psilocybin or LSD had a history of approximately 50 percent fewer incidences of inflicting violence on an intimate partner compared to those who did not have these psychedelic experiences. The male psychedelic users also reported better emotion regulation when compared to men with no history of psychedelic use.[41] In my personal experience as well, those who have used psilocybin tend to be kinder, more patient, better listeners, and less prone to react negatively when provoked.

OPIOID USE DISORDER

In a meta-analysis of responses from 214,505 participants to the National Survey on Drug Use and Health (2015–2019), only past use of psilocybin—not any other psychedelic—was associated with a reduction in opioid use disorder.[42] Notably, eight clinical studies are ongoing to treat opioid addiction, but the results have not yet been reported.

Many more clinical studies are in the queue. The range in scale and cost is great. A Phase I study that primarily tests safety and initial efficacy may have as few as 20 patients and cost around a million dollars. A Phase III study may include thousands of patients and cost hundreds of millions of dollars. Without a profit incentive, many studies on psychedelics will never get past Phase I or II. Peer-reviewed Phase III studies with positive outcomes are necessary for any approved new drug to get to market. Good clinical studies with psilocybin (and related tryptamines) might prove to the regulators within government and the medical community to allow approval for helping people suffering from many maladies. Ironically, so many users of psilocybin mushrooms have benefitted—and continue to do so—without regulators sanctioning use. This is the long dance between the development of approved medicines from natural products.

Patents for psychedelic products sound problematic to some users with fears that pharma could steal their right to dose. Psilocybin mushrooms can be easily grown by most people, and patents cannot stop your personal use. Indigenous use is and should always be exempt—in my opinion—as well as religious use by anyone. Many who object to patents do not fully understand their ultimate merit to society. Patents are commercial instruments in a world where capital is an unavoidable part of the narrative. Patents fundamentally address a critical issue: If the government won't step up to provide universal access to psilocybin, then I think we need to increase access through regulated businesses.

Proving the medical and societal benefits of psilocybin, and psychedelics, at scale requires expensive research that can be funded by prospective investors. Most would not risk their money unless patents are already in place to protect intellectual property. Ultimately, patents enrich the Commons after temporarily rewarding those with novel ideas and their risk-taking fiscal backers. Once patents expire—and they all do—researchers' and inventors' discoveries are free for anyone to exploit for profit in the public domain.

In my opinion, governments should fund studies and offer free access to psilocybin in all its forms for the benefits of citizen health and happiness, at no cost. The reduction in depression, addiction, crime, and potentially dementia and other neurodegenerative disorders could save hundreds of billions of dollars and elevate society to a higher level of existence. As obvious as this is, it seems unlikely that we can depend on governments to take such bold, reasonable steps, given our long history of dysfunctionality. Nevertheless, governments can help steward access to psilocybin by deploying thoughtful regulations and guidelines for allowing commercialization.

A handful of the potent *Psilocybe cyanescens*, the Wavy Cap.

Part 2

Identification Guide to Psilocybin Mushrooms

Chapter 7

The Tricksters

This second half of the book is a field guide to the most common psilocybin mushrooms you are likely to encounter, along with their close look-alikes—what I call the tricksters. I list the *poisonous* look-alikes first, followed by the nonpoisonous ones, and I suggest that you familiarize yourself with these important species before turning to the psilocybin-active species described in Chapter 8. After touring these look-alike species and learning which ones are psilocybin-inactive and rather *poisonous*, you will be better prepared to identify the psilocybin-active species.

Like most of you, when I go looking for psilocybin mushrooms, I encounter many types of mushrooms, some of which look enticingly similar to psilocybin species. I compiled the following list based on my frequent encounters with these species, but, of course, those of you not living in my bioregion will encounter unique species. That said, most of these tricksters are very common and can be found across wide swaths of the world.

Many of these tricksters are also indicator species, meaning they often grow in close proximity to psilocybin mushrooms. Since humans are likely to continue converting woodland habitats into suburban and grassland (lawn) habitats, I've listed both woodland and grassland trickster species here as they can co-occur, particularly in suburban areas where wood chips are used for landscaping that closely interfaces with manicured lawns.

The essential burden on you is the ability to recognize both poisonous and psilocybin mushrooms. To not know the poisonous species puts you, your friends, and your family at risk. Mistakes in mushroom identification are common—a few can be deadly. Remember, do not consume a mushroom unless you are sure of its identification! And be careful when consuming mushrooms that were picked by others where different species can be indistinguishable, especially if powdered. Ultimately, you are responsible for your own health and safety. The adage "When in doubt, throw it out" has prevented many accidental poisonings, so keep this in mind. Or, as citizen scientists employ this better adage: "When in doubt, find out."

Welcome to the marvels of mycology. There are so many fantastic and fun species for you to discover. Let's get started.

Many mushroom species resemble psilocybin-active ones. These tricksters, *Entoloma* (*Leptonia*) *serrulata* (top left), *Mycena amicta* (top right), and *Hypholoma dispersum* (bottom) resemble some Psilocybe mushrooms but are not psilocybin-active. They often fool beginners.

The Poisonous Tricksters

Galerina marginata, the Funeral Bell. This rusty brown–spored mushroom is DEADLY. In many books, this species is listed as *Galerina autumnalis*, a synonym.

Galerina marginata. It contains the DEADLY cyclopeptides (amanitins) of the Destroying Angels, *Amanita phalloides, Amanita bisporigera*, and *Amanita virosa*. Grows on dead or dying trees, on logs, and from wood chips.

The rusty brown–spored *Galerina castaneipes*. Another DEADLY Galerina, this one is often mistaken for *Galerina marginata,* but both species can kill you if you ingest enough. Grows on dead or dying trees, on logs, and from wood chips.

The *Omphalotus olivascens* complex is TOXIC, containing a sesquiterpene compound known as illudin S, which is also tumoricidal. This rusty brown–spored mushroom is commonly called the Jack O'Lantern Mushroom because it glows in the dark. Historically this species was listed in field guides as *Omphalotus illudens* or *Omphalotus olearius*. These species can be confused with *Gymnopilus* or, if small, with *Galerina* species. Some varieties of *Gymnopilus junonius* have been reported to contain neurotoxic oligoisoprenoids. All grow at bases of dead or dying trees, on logs, and from wood chips.

Pholiotina rugosa (= *Pholiotina filaris*) is a DEADLY, rusty brown–spored mushroom with a distinctive but very fragile membranous annulus that tends to form midway up the stem. Rain, wind, or handling can easily remove this distinctive ring. It can grow in grasses or in areas mulched with wood chips. I have found it fruiting in a potted Douglas fir tree (see page 73).

Hypholoma fasciculare, the Sulphur Tuft or Green Gilled Clustered Woodlover, is a very common TOXIC mushroom. Grows on dead or dying trees, on logs, and from wood chips.

The Nonpoisonous, Psilocybin-Inactive Tricksters

The rusty brown–spored *Conocybe tenera* group is a very common nonpoisonous mushroom growing in grassy areas.

Deconica coprophila is a common species with purple brown to black spores that grows on dung from many animals, both wild and domesticated. It does not contain psilocybin. The hemispheric cap is typical.

I have found the purple brown–spored *Deconica angustispora* on deer and elk dung. Psilocybin has not *yet* been detected in this species. I find this surprising, as it closely resembles many psilocybin-active, dung-dwelling species, although DNA analysis places it into *Deconica*.

Deconica montana is a purple brown–spored mushroom that typically grows with mosses. It does not contain psilocybin.

Hypholoma dispersum is a purple brown–spored, typically bluntly cone-shaped mushroom that does not contain psilocybin. This species is very similar to conic *Psilocybe* species, like *Psilocybe pelliculosa*, but lacks a separable gelatinous pellicle. Grows from wood chips.

Hypholoma capnoides is a purple brown–spored, edible mushroom closely related to the poisonous *Hypholoma fasciculare* (see page 123). Grows on dead or dying trees, on logs, and from wood chips.

Hypholoma ericaeum is a purple brown–spored, nonpoisonous mushroom that grows on wood chips or wet soils enriched with wood debris.

Leratiomyces ceres is a not-so-good edible mushroom that has purple brown spores. Grows on wood chips.

Panaeolus antillarum complex is a psilocybin-inactive, purplish black–spored, whitish mushroom that thrives on horse dung and lesser so on the dung of other animals. Not known to be poisonous.

Panaeolus rickenii (right; = *Panaeolus acuminatus*) produces dark purplish brown to black spores and grows in grassy areas, especially dung-enriched pastures. It is not known to be poisonous. The other mushrooms are classic *Panaeolus papilionaceus sensu lato* (left; = *Panaeolus sphinctrinus* and *Panaeolus campanulatus*), which are also not poisonous.

The dark brown–spored (*not* purple brown–spored) psilocybin-inactive *Panaeolus foenisecii*, a Haymaker's Panaeolus, flourishes in lawns, fields, and grasslands. It is very common and not known to be poisonous or psilocybin containing. See also the psilocybin-active *Panaeolus cinctulus* (page 205).

Panaeolus foenisecii fades in color as it dries, a feature known as hygrophanous. Psilocybin-inactive.

Panaeolus papilionaceus is a dung-loving species with purplish black to black spores that has had many names, including *Panaeolus sphinctrinus* and *Panaeolus campanulatus*. Not known to be poisonous; psilocybin-inactive. Note that the denticulate triangular veil remnants on the cap edge are barely visible. These often wash off from rains or in the wind. See also page 39.

Panaeolus semiovatus is a regal, statuesque species growing on dung, particularly horse dung. It produces purplish black spores and is not known to be poisonous. Psilocybin-inactive.

Protostropharia dorsipora is very similar to *Protostropharia semiglobata* and *Protostropharia luteonitens*, all of which grow on dung or in manure-enriched soils. All have yellow- to yellowish brown–toned caps and purplish black spores. Psilocybin-inactive.

Protostropharia luteonitens grows on dung. Psilocybin-inactive.

Protostropharia semiglobata grows on dung. Psilocybin-inactive.

Protostropharia semiglobata (left, psilocybin-inactive) and *Psilocybe semilanceata* (right, psilocybin-active) both grow in dung-enriched grasslands. *Protostropharia semiglobata* is an indicator species for psilocybin mushrooms like the psilocybin-active *Psilocybe semilanceata* (Liberty Cap), which often thrives nearby. *Protostropharia semiglobata* is covered with a glutinous sheath, making it slippery and difficult to pick when moist. It is not poisonous. Both have purplish brown to black spores.

Protostropharia semigloboides has purplish brown to purplish black spores and grows from conifer-enriched soils. I found these specimens on a trail in an old-growth Douglas fir forest, covered with a glutinous veil that makes the mushroom slippery to pick, a feature shared with the dung-loving *Protostropharia semiglobata*. Neither are known to be poisonous or to contain psilocybin.

Psathyrella longipes group has blackish spores and typically grows in very wet lowlands and saturated soil enriched with wood debris. Not known to be poisonous. Psilocybin-inactive.

Psathyrella piluliformis, and similar species in the *Psathyrella hydrophila* group, grow in soggy habitats rich in leaf litter and wood debris. Not known to be poisonous. Psilocybin-inactive.

Psilocybe atrobrunnea (= *Psilocybe turficola*) grows in boggy areas lush with sphagnum mosses. Not known to be poisonous. It has *not* been found *yet* to contain psilocybin—hence its placement here. However, genomic analysis places the species in a psilocybin-active clade. I also describe this species in Chapter 8 because, although the varieties analyzed to date do not contain psilocybin, it would not be surprising to find some that do. (See also *Hypholoma ericaeum, Psilocybe pelliculosa, and Psilocybe silvatica* on pages 125, 180, and 187.)

Chapter 8

The Psilocybin-Active Species

Now that you have been introduced to some of the dangerous look-alikes, let's dive into the psilocybin-active species. The world is just entering into a new dawn of psilocybin science that promises many more discoveries. More than 220 species of psilocybin mushrooms are known, with several more species being discovered each year. Not all have been analyzed, but teams of scientists are evaluating them for their psilocybin content.[1] As citizen scientists, we all can significantly contribute to expanding the field of knowledge.

Most psilocybin mushrooms have black to purplish brown to brown spores. No known psilocybin-active species produce greenish, yellowish, or whitish spores—a fact that I find to be quite remarkable. It seems that horizontal gene transfer via bacteria would have crossed over to these latter genera, carrying the psilocybin gene-making machinery with them. Perhaps psilocybin-active mushrooms with greenish, yellowish, or whitish spores are out there, but we have yet to find them. Remember that we have only identified approximately 10 percent of the estimated fungal species that form mushrooms. Many more may lurk, hidden from view.

I have prioritized the mushrooms with purple brown to black spores in the genera *Psilocybe* and *Panaeolus* followed by the brown-spored mushrooms in the genera *Conocybula, Gymnopilus, Inocybe,* and the pinkish- to salmon brown–spored *Pluteus* species. As I have repeatedly emphasized, mushrooms with purple brown to black spores are a safer group to explore. Those producing rusty brown to cinnamon brown spores are inherently problematic because there is a mixture of psilocybin-active and poisonous mushrooms within these genera.

More psilocybin-active species are contained within the genus *Psilocybe* than any other, followed by *Panaeolus*. *Psilocybe* species typically have purplish brown to violet brown spores, whereas *Panaeolus* hosts mostly black-spored mushroom species. When more spores are deposited en masse, the spore print becomes visually darker, so the difference between dark purplish brown and black can be subtle. In this case, the difference between a mushroom spore print looking black versus purplish brown is generally inconsequential.

In the brown-spored genera of *Gymnopilus, Pholiotina* (sensu lato), and *Inocybe*, there are

This psilocybin-active *Psilocybe hoogshagenii* is a classic Psilocybe.

both psilocybin-active *and* poisonous species. The brown-spored psilocybin-active species are in the genera *Conocybula* (a new genus split off from *Pholiotina*), *Gymnopilus, Inocybe,* and *Pluteus*. However, in the genus *Pholiotina* is *Pholiotina rugosa*, also known as *Pholiotina filaris* (see photo on page 123), a deadly species having similar amatoxins as those found in the white-spored Destroying Angel, *Amanita phalloides*, and the brown-spored Autumnal Galerina, *Galerina marginata*, which has also been called *Galerina autumnalis*. An interesting fact: no psilocybin mushroom species has been found *so far* that produces both psilocybin and these deadly amatoxins. There is one *Galerina* species known to contain psilocybin, *Galerina steglichii*.[2] However, out of an abundance of caution, I have not described this exceptionally rare *Galerina* because of a strong potential for misidentification. Since there is ample room for confusion, I suggest people avoid experimenting with non-purple brown-spored species in general. The purple brown- to black-spored species are much safer.

A mistake in identification can be—and has been—deadly. Practice an extraordinary abundance of caution before ingesting any mushroom. Be sure you are *not* ingesting a poisonous species and, if you choose to ingest a psilocybin mushroom, do so safely and with respect and employ safeguards to ensure a positive, healthy experience. Please memorize the poisonous species and do not experiment with the lighter brown-spored (i.e., not purple brown- not black-spored) genera. This is why I describe in detail how to make a spore print on pages 74–75. I provide information on these species for academic reasons and do not recommend that amateurs ingest any psilocybin mushroom species that has brown spores because of the potential for misidentification. Lastly, an obvious but necessary warning: if you are color blind or have difficulty discerning shades of brown, identification of suspected wild psilocybin mushroom species could be a dangerous endeavor!

In the entries that follow, each species is organized by a description of the mushroom's **cap, gills**, and **stem**; **microscopic features** including spore size, shape, color, and the shape and sizes of cystidia. Since the vast majority of species have 4-spored basidia, I do not remark on basidia unless they are 2- or 3-spored. I also list **habit, habitat, and distribution,** although the distribution of many species is expanding as myconauts are spreading them. Climate change is also a factor. In the **comments**, I discuss potency, taxonomy, and other information.

Classic Liberty Cap (*Psilocybe semilanceata*), a common, potent, typically non-bluing , purple brown–spored psilocybin-active grassland species.

The Genus *Psilocybe*

The genus *Psilocybe*, meaning "smooth head" or "bald head" is characterized by saprophytic, purplish brown–spored terrestrial mushrooms, small to medium in size, with attached gills. (Purplish brown spores can look nearly black when they heavily amass in a spore print.) Their smooth spores have germ pores at one end.[3] The genus contains numerous species that are psilocybin-active, many of which bruise bluish upon injury (from handling), natural growth, or exposure to weather.

The *Psilocybe* lineage can be traced back 65 to 67 million years, and it diversified into two clades with different gene order patterns 57 million years ago. Based on mutation rates (i.e., molecular clock models), Bradshaw et al. suggested that psilocybin synthesis likely spread to other mushrooms via horizontal gene transfer somewhere between 40 and 9 million years ago.[4] Remarkably, *Psilocybe* mushrooms far predate us hominids, who first appeared 7 to 6 million years ago. Modern humans evolved much later, with the earliest records from about 360,000 years ago and later estimates about 160,000 years ago. Mushrooms have had their forms far longer than we, as *Homo sapiens*, have had ours. They are ancient organisms. We are recent.

The genus *Psilocybe* has undergone many revisions in interpretation by mycologists over the years. Elias Fries first used the name *Psilocybe* for a group of species within the large genus *Agaricus* in 1821.[5] In 1871, the name was used by Paul Kummer to describe a genus, but it included many species that are no longer considered to be *Psilocybe* as defined currently.

In 1946, Rolf Singer and Alexander H. Smith established the family Strophariaceae, which included *Psilocybe* along with other member genera that have also undergone numerous revisions. At one time the family included eighteen genera.[6] The boundaries between the genera *Psilocybe* and *Stropharia* have been debated as the science has progressed. For example, Franklin Sumner Earle first described the species *Stropharia cubensis* from Cuba in 1906.[7] In 1949, Rolf Singer transferred this species to *Psilocybe*, renaming it *Psilocybe cubensis* (Earle) Singer, and later contributed to a monograph on *Psilocybe* published in *Mycologia*.[8] In 1984, Gastón Guzmán published a monograph on the species known worldwide, entitled *The Genus Psilocybe*,[9] in which he added and combined many species, and in 1995 he supplemented it with an update. That same year, Machiel Noordeloos placed many of the psilocybin-inactive species in the genus *Stropharia* into *Psilocybe*, following an expanded definition proposed by Singer and Smith.[10] I followed this definition when writing my book *Psilocybin Mushrooms of the World*.

The genus *Psilocybe sensu lato* (meaning "in the broadest historical interpretation") contained two major groups of species: one that was psilocybin-active and debatably classified in the separate Hymenogastraceae family and the other psilocybin-inactive and classified in the Strophariaceae family. Prior to molecular-based taxonomy, Strophariaceae usually included *Hypholoma* (formerly *Naematoloma*), *Stropharia, Deconica*, and *Pholiota*.[11,12] In 2007, however, researchers noted that the traditional *Psilocybe* species are naturally divided into two groups and should be placed into two genera, the psilocybin-active and the psilocybin-inactive species.[13] Species newly classified within the genus *Deconica* include *Psilocybe montana*, which was formerly designated as the type species (i.e., a species that best represents most of the closest interrelated mushroom species) of the genus *Psilocybe* by some authors; the psilocybin-inactive *Psilocybe merdaria*, which was considered to be the type species by other authors; and *Psilocybe coprophila*, a likewise psilocybin-inactive species, among others. The new type species of the genus *Deconica* is *Deconica montana*. *Deconica* species are often petite, squat mushrooms with gills bluntly attached (adnate) to the stems and lacking psilocybin. They often co-occur with psilocybin-active species (see page 125).

The now representative or "conserved" type species of the genus *Psilocybe* is *Psilocybe semilanceata*. The combined efforts of many mycologists involving DNA-based molecular analyses followed by nomenclatural conservation to protect the name *Psilocybe* has stabilized which species can be called *Psilocybe*, as ratified by the International Botanical Congress in 2011.[14] Now the genus *Psilocybe* encompasses all the psilocybin-active mushroom species with a notable exception: *Psilocybe atrobrunnea* (=*Psilocybe fuscofulva*), in which psilocybin has not yet been detected, despite DNA evidence placing it into a clade of psilocybin species. Borovička et al. noted "biosynthesis of these alkaloids was lost in *Psilocybe atrobrunnea*."[15] I list this species here as a *possible* psilocybin mushroom because more analyses are needed and some phenotypes of this species *may* produce psilocybin.

Hence, I follow this new definition of *Psilocybe sensu stricto* (meaning "in the narrowest, most precise interpretation"), within the modern definition of the Strophariaceae and Hymenogastraceae families. In my previous book, *Psilocybin Mushrooms of the World*, I followed the broadest interpretation, a *sensu lato* interpretation, of the genus *Psilocybe*, which included many psilocybin-inactive species scattered among several genera, particularly those known then and now in the genus *Stropharia*. With modern DNA tools, we now have a much clearer picture of their evolutionary heritage. Taxonomy is a science that is constantly in a state of flux and refinement.

Presently, of approximately 220 psilocybin-active species identified thus far, 168 species are recognized as *Psilocybe* species. However, the number of species is changing because many names are being synonymized while new species are being discovered. Work by Alexander Bradshaw, Bryn Dentinger, and their teams[16] have helped to identify the clades (natural groups) within the *Psilocybe* genus, clarifying phylogenetic trees of association based on DNA.

Following are descriptions of the Psilocybes and the psilocybin-active *Panaeolus* species, then other psilocybin mushrooms scattered among several brown-spored genera. Psilocybin-active species thrive on every continent of this planet except Antarctica. In the following descriptions, note that the precise length of the stems, breadth of the cap, and other minor features are greatly influenced by environmental conditions. These measurements are generalized here for most of the phenotypes you will likely encounter, so do not expect these ranges to strictly match your specimens. I also used the broadest ranges reported for the microscopic features.

Psilocybe allenii is a potent psilocybin mushroom that appears to have originated in the Bay Area—Oakland and San Francisco—but is rapidly spreading. Most all psilocybin mushrooms grow very rapidly, maturing in a week. These are adolescent forms, a relatively young patch of *Psilocybe allenii* that likely first became visible as primordia 3 to 4 days before.

Psilocybe allenii BOROVIČKA, ROCKEFELLER & P. G. WERNER

CAP: 1.5–7(9) cm in diameter, hemispheric, convex to broadly convex, expanding to plane. Lacking an umbo at the cap center. Margin even, incurved when young, flaring in age, often with eroded edges, sometimes wavy. Cap edge translucent striate when moist. Surface smooth, viscid when moist, often with a separable gelatinous pellicle, hygrophanous, light orangish brown to caramel when wet, fading to a light yellowish brown upon drying. Flesh bruising bluish or with bluish stained zones naturally from environmental stress. **GILLS:** Attached, adnate to sinuate, cream to pale light brown when young, darkening with maturity from purplish brown spores, eventually becoming dark purple at maturity. Gill margins fringed whitish from cheilocystidia. **STEM:** 40–90 mm long by 2–7 mm thick, even, hollow, cartilaginous, becoming pruinose toward the apex, swollen toward the base, with thick whitish rhizomorphs. Surface glabrous to fibrillose, whitish when young and strongly bruising bluish. Becoming tinged with yellowish tones in maturity, bruising bluish. **MICROSCOPIC FEATURES:** Spores ellipsoid with a germ pore and relatively thick wall (0.8–1 μm), brownish with yellowish tones in 5 percent KOH, 12–13 by 6.5–7.5 μm. Cheilocystidia abundant, variable, attenuating to clavate-mucronate, lageniform, rarely forking, infrequently fusiform to fusiform, hyaline, thin-walled. Pleurocystidia common, narrowly to broadly clavate mucronate, hyaline, thin-walled, mostly 25–35 μm by 9–14 μm. Cystidia can be encrusted at apex.

Psilocybe allenii can be easily naturalized in your backyard. This is an outdoor alder wood-chip bed of *Psilocybe allenii* fruiting 6 months after inoculation, but typically for only 1 or at most 2 years thereafter. Most patches like these benefit from an annual renewal of fresh hardwood chips.

Psilocybe allenii produces abundant, radiating rhizomorphs at the base of the stem. When the stem is trimmed, the butt can be used to create new outdoor patches.

HABIT, HABITAT, AND DISTRIBUTION: In groups or clusters, sometimes appearing in large numbers from late September through December. Typically, a western North American coastal species and reported up to 100 miles inland, growing from Los Angeles to Seattle. Mostly found in wood chips used for landscaping, to gregarious, sometimes cespitose, growing on woody debris, usually on wood chips (Monterey pine, Monterey cypress, eucalyptus, Douglas fir) and primary saprophyte, temporarily present by decaying wood chips, and absent unless wood chips are replenished every few years. This species is migrating to other regions of the world.

COMMENTS: A moderately potent species and, given its strong bluing reaction, *Psilocybe allenii* is closely related to *Psilocybe azurescens, Psilocybe cyanescens, Psilocybe ovoideocystidiata,* and *Psilocybe subaeruginosa*—all of which thrive in wood chips and wood mulch commonly used for landscaping and in tree nurseries, among roses, rhododendrons, and other woody shrubs. In the early 1980s, I found this species at the Oakland Museum just prior to giving a talk to the Mycological Society of San Francisco. Before and after my talk, dozens of myconauts collected specimens and thereupon launched this species into new locations. In the years following, numerous reports of its occurrence became commonplace, as much as if not more than *Psilocybe cyanescens,* with which it competes. The typically even cap margin and gregarious nature makes it most visually similar to *Psilocybe cyanescens* but without the classic sine wave undulating margin. When cloning specimens, I was surprised that one culture grew out in spiral waves. (See page 8, Figure 8 of *Mycelium Running: How Mushrooms Can Help Save the World.*) Many people are now cultivating this species, so its eco-geographical range is rapidly expanding. This species is named in honor of John W. Allen for his lifetime dedication to exploring psilocybin mushrooms.

Like many psilocybin-rich mushrooms, *Psilocybe allenii* turns blue after freezing.

Psilocybe atrobrunnea is a possible psilocybin species. Its placement, based on DNA analyses, is directly within clusters of species well known for their psilocybin activity. Its production of psilocybin may be intermittent or strain dependent. We don't know.

Psilocybe atrobrunnea (LASCH) GILLET
= *Psilocybe turficola* J. FAVRE
= *Psilocybe fuscofulva* PECK

CAP: 2–4(6) cm broad. Obtusely conic, expanding to campanulate to convex, broadly convex to nearly plane in age. Reddish brown, darkening in age and, with spore maturity, strongly hygrophanous, fading to pale reddish brown in drying. Surface smooth, translucent striate near the margin, viscid when wet with a separable gelatinous pellicle. Margin incurved at first, soon expanding, and ringed with tufts of whitish veil remnants. **GILLS:** Attachment adnate to adnexed, dull cinnamon brown to dark purplish brown at maturity, with edges fringed whitish. **STEM:** 80–180 mm long by 3–5(6) mm thick. Equal, tough, sometimes flexuous, cured and enlarging toward the base, which is often tufted with whitish mycelium, sometimes with faint bluish tones. Reddish to blackish underneath sheaths of appressed whitish, fibrillose patches but tending to be pruinose near the apex. **MICROSCOPIC FEATURES:** Spores dark purplish brown in deposit, subellipsoid, 9–12(14) by 5–7(9) µm. Pleurocystidia absent. Cheilocystidia lageniform or fusoid ventricose with a tapering neck, 18–36 by 4–7 µm, and 1.5–2.5 µm at apex.

HABIT, HABITAT, AND DISTRIBUTION: Numerous to scattered, typically associated with sphagnum mosses, around bogs in both coniferous and deciduous woodlands. Fruiting from August through November. Reported across the northern United States, western and southern Canada, and across Europe and Russia. Probably more widely distributed.

COMMENTS: The names *Psilocybe atrobrunnea* (Lasch: Fr.) Gillet, *Psilocybe turficola* J. Favre,[17] and *Psilocybe fuscofulva* Peck have been considered synonyms.[18] This mushroom is a classic *Psilocybe* similar in appearance to psilocybin-active *Psilocybe pelliculosa, Psilocybe silvatica*, and *Psilocybe medullosa*. Although psilocybin has not yet been detected in a few analyses of *Psilocybe atrobrunnea*, its close genetic relationship to other psilocybin-active species strongly suggests that this species could be active, albeit at low levels or intermittently. I would not be surprised if some strains are psilocybin-active while other strains are not. Dentinger suggested that this is one of the earliest diverging *Psilocybe* species, making it the progenitor of many species that are psilocybin-active and originating many tens of millions of years ago.[19] Why it may be inactive today—or intermittently active—is a mystery. A collection from Vancouver, British Columbia, showed slight bluish tones at the base of the stem.[20] See also *Hypholoma dispersum* and *Hypholoma ericaeum*.

Adolescent forms of *Psilocybe aztecorum*, a potent mushroom found in southern Mexico and Quebec, Canada.

Psilocybe aztecorum gills become grayish black with spore production.

Caps of mature specimens of *Psilocybe aztecorum* are often darker than younger forms as the purple brown spores mature on the underlying gills. The variability from younger to more mature is typical of many *Psilocybe* species. Note rhizomorphs attached to stem bases. These stem butts can be transplanted to create satellite colonies.

Psilocybe aztecorum R. HEIM = *Psilocybe quebecensis* OLA'H & R. HEIM

Common Names: Niños, Los Niñitos in Spanish, Apipltzin (= little children) in Nahuatl

CAP: 1.5–3.5 cm broad. Bluntly conic soon convex, expanding to broadly convex, plane when mature. Surface viscid when moist, smooth, translucent striate along the margin. Golden to caramel brown when young, darkening to dark chestnut brown in age, sometimes dark gray brown with greenish tinges, strongly hygrophanous, drying straw yellow, and easily bruising bluish. Cap margin is typically even. **GILLS:** Attachment adnate to adnexed, light purplish gray to dark purplish brown, fringed whitish on edges. **STEM:** 25–95 mm long by 3–4 mm thick. Equal, thickening toward the base and apex, straight to flexuous, pruinose above and silky-fibrillose below. White rhizomorphs radiate from the base of the stem. Whitish to grayish, easily bruising bluish. Partial veil cobwebby, white, soon disappearing, and occasionally leaving an annular zone in the upper reaches of the stem. **MICROSCOPIC FEATURES:** Spores dark purple brown to violet black, 10.5–14 by 7–9 µm, ellipsoid. Basidia 1-, 2-, 3-, or 4-spored. Pleurocystidia few, scattered, same as cheilocystidia. Cheilocystidia 20–45 by 5–8 µm, fusoid shaped, with a long thin neck 6–11 by 1.5–2.5 µm.

HABIT, HABITAT, AND DISTRIBUTION: Found in soils rich in wood debris, rarely on pine cones, in open woods of *Pinus hartwegii*, *Pinus montezumae* and fir, and *Abies religiosa* interspersed with grasses. Fruiting from August through October in the high mountains of central Mexico 6500–13,000 feet (2000–4000 m) above sea level. Also known from the White Mountains of Arizona, Vermont, possibly Montana, and Quebec and Nova Scotia. Probably more widely distributed.

COMMENTS: This woodland species shares many characteristics with the more temperate *Psilocybe* species such as *Psilocybe cyanescens, Psilocybe ovoideocystidiata*, and other chestnut brown to yellowish brown to caramel-colored species, especially when young. *Psilocybe quebecensis* is a synonym. Given its prolific rhizomorphs adorning the stem bases, this is a prime candidate for creating naturalized beds by transferring stem butts.

Psilocybe azurescens STAMETS & GARTZ

Common Names: Azs, A-zs, Azzies, Flying Saucers

CAP: 3–10 cm broad. Bluntly conic, maturing to broadly convex, eventually flattening with maturity. Surface smooth, viscid when moist, covered by a gelatinous pellicle, often not separable. Cap margins characteristically even, not wavy, with translucent striate lines. Can have tints of azure blue mixed within the brown caps as they mature. Obtuse broad umbo characteristic, most evident as the caps flatten. Color deep chestnut to caramel, often with dark blue or bluish black zones, especially where there are lesions or bruising. Caps hygrophanous; brown fades to straw yellow or light dingy brown as they dry out, with the bluing reaction becoming more pronounced. Caps typically dark brown when fully dried. Bruising deep blue to bluish black when damaged. Cap thickness (context) at center 1–10 mm in cross section, cottony and whitish at the stipe–pileus junction. Flesh whitish to pallid brown under the umbo, but thinning toward cap margins, soon azure blue, then darkening to indigo to black with damage or age. **GILLS:** Attachment ascending, sinuate to adnate, medium brown in young specimens, darkening with spore maturity, with two tiers of short intermediate gills. Gills mottled from variable regions of spore maturity. Gill edges have a whitish fringe. Spore prints dark purplish brown to purplish black in mass. **STEM:** 50–200 mm long by 2–6 mm thick, silky white, often tinged with azure blue, but dingy brown from the base or in age. Hollow at maturity but encompassed by lengthwise arranged, twisted, stringy fibrous strands of tissue. Base thickening downward, often curved, adorned with white aerial fuzzy tufts of mycelium, from which white rhizomorphs emanate. These rhizomorphs tenaciously attach the stems to wood chips or

The caramel-colored *Psilocybe azurescens*, aka Azzies, naturalized near Sebastopol, California, and fruiting from oak chips. High in psilocybin content, this mushroom in high doses is also associated with bouts of temporary paralysis and loss of muscle control and coordination in some people. The compound(s) causing this remains a mystery.

thatch. Partial veil white, cobwebby, well developed, often leaving a fibrillose annular ring in the upper regions of the stem that becomes dusted with purplish brown spores. **MICROSCOPIC FEATURES:** Spores dark purplish to dark lilac brown in deposit, 12–13.5 by 6.5–8.0 µm, ellipsoid, brown under the microscope. Pleurocystidia abundant, fusoid-ventricose, tapering to a narrow but short neck, bluntly papillate, 23–35 by 9–10 µm. Cheilocystidia nearly identical to pleurocystidia, 23–28 by 6.5–8.0 µm.

HABIT, HABITAT, AND DISTRIBUTION: Gregarious to clustered, occasionally solitary, on deciduous wood chips and/or in sandy soils rich in lignicolous debris infused with grasses. This species naturally grows just north and south of the Columbia River along the coastlines of northern Oregon near Astoria to the first few miles northward into southwestern Washington. *Psilocybe azurescens* favors the few miles of the beach-land interface. This species is associated with dune grasses, especially *Ammophila arenaria* and *Ammophila breviligulata*, which the U.S. Army Corps of Engineers planted for erosion control along the coastline where the Columbia empties into the Pacific. Over the ensuing decades, several thousand acres of new sandy habitats emerged, interspersed with shore pines, Sitka spruce, and speckled throughout with dune grasses. *Psilocybe azurescens* proliferates in this newly expanded habitat.

Where is the ancestral origin of this species? It could have drifted down the Columbia River from log and wood debris—perhaps from a beaver dam that burst. Or it may have colonized this area from afar, from mycelium-impregnated boards in any of the more than two thousand ships that have broken apart trying to navigate

Psilocybe azurescens is a comparatively large and potent Psilocybe, endemic to the shorelands near where the Columbia River flows into the Pacific Ocean between Oregon and Washington. It thrives in sandy soils and decomposing mixtures of grass thatch replete with broken fragments of fir and alders.

through the shallows of the Columbia River Bar. We may never know. With thousands of people collecting this mushroom species and transplanting its mycelium, new natural habitats are being created. Although not native to Canada and Europe, outdoor patches have been established there that could become launching sites for this species to spread and acclimate in those regions.

Please be forewarned: "No Mushroom Picking" signs are conspicuously placed at several Oregon and Washington state parks, and law enforcement vigorously enforces the rule. Fines for picking mushrooms have been a source of income for the local municipalities. A game-like cat-and-mouse dynamic is at play. Photographing—but not picking—the mushrooms is legal but will likely draw attention (see page 60).

Generating an extensive, dense, and tenaciously ropey mycelial mat, *Psilocybe azurescens* whitens the wood it colonizes. Fruitlings build in October and continue well after the first frost, maximizing in late October through November, fruiting into late December, and dwindling in early January. This species is easy to naturalize

Psilocybe azurescens is known for its size, blunted umbo, chestnut brown color, and strongly bluing whitish stem.

in established outdoor beds if temperatures hovering around freezing occur occasionally during the fall. A cold snap triggers fruitings, which is different than its close cousin, *Psilocybe cyanescens*, which typically has a fruiting window several weeks earlier and is harmed by freezing. Both species can co-occur in the same habitat, although *Psilocybe cyanescens* has a broader natural distribution than the more geographically limited original natural ecozone of *Psilocybe azurescens*. *Psilocybe azurescens* is very close genetically and in appearance to *Psilocybe subaeruginosa*, which is one of the most potent, dominant *Psilocybe* species in Australia.

COMMENTS: A species tolerant of low temperatures, *Psilocybe azurescens* is one of the most potent species in the world, consistently having more than 2 percent psilocybin and psilocin.[21] The analysis was done 6 months after harvesting with the specimen stored at room temperature after drying in a heated food dehydrator, which means the potency when freshly harvested is likely much more. High in psilocybin, psilocin, and baeocystin, this species exhibits one of the strongest bluing reactions I have seen. The flesh can become indigo to black where damaged or post freezing. The silky white stem, caramel-colored cap, relatively large stature, even cap margin, and broad but pronounced umbo make it distinct.

At high doses, some report temporary lack of coordination, specifically loss of muscle control and motor skills. Unless forewarned, this can be alarming to unsuspecting consumers. Some describe this as temporary paralysis. The syndrome has been given the name "wood lover's paralysis" or "wood-chip mushroom paralysis." Those afflicted generally fully recover in a day. The biggest concerns I have are the attendant risks during intense journeying: hypothermia, inability to move from danger, incoordination, anxiety, and panic.

I had this species analyzed for endotoxins from bacteria at a medical facility in Seattle, thinking that the black necrotic lesions could be harboring harmful toxic bacteria. No pathogenic bacteria were found. Another potentially new toxin or analgesic may be present. This species is likely to contain aeruginascin. Sherwood et al. found that aeruginascin, a psilocybin-related tryptamine of unknown toxicity, could be converted to bufotenine, a psychoactive tryptamine (5-hydroxy-N,N-dimethyltryptamine) also found in toad venom.[22] Chadeayne reported that aeruginascin can bind to the same 5HT2A/B receptors as psilocybin as well as 5HT3 receptors, and aeruginascin likely undergoes a similar hydrolysis as psilocybin when it is converted to psilocin.[23] Jochen Gartz first described and Jensen et al. reported that aeruginascin's effects were euphoric, with no mention of paralysis.[24,25] The loss of muscle control remains a mystery. I do not experience paralysis with *Psilocybe cyanescens* as I have with *Psilocybe azurescens*, although some people have. In any event, I have little motivation to move around under high dosages of either mushroom.

Today, I avoid ingesting this species until its chemistry is better understood. When I have ingested it in the past, I experienced a loss of coordination and weakness in the limbs, but not paralysis. I preferred to lie down, endure, and think, and had no motivation to move. The experience was more physical than cerebral and did not impart the visuals I often find fascinating and intriguing. I felt like a liquid being internally, oceanic, being tugged by the tides of the universe. Notably, some people do not report these symptoms, which may be dose-dependent or idiosyncratic.

Psilocybe baeocystis is a potent psilocybin mushroom. Note the bluing reaction that formed directly after harvesting these. Also, the dark cap color (when fresh) and the convoluted margins are typical.

Psilocybe baeocystis SINGER & A.H. SMITH

Common Name: Baeos

CAP: 1.5–5.5 cm broad. Bluntly conic to convex, infrequently expanding to near plane only in extreme age and, if so, with an irregular cap margin. Margin incurved at first and distinctly undulated when convex; translucent striate and often tinted greenish. Dark olive brown to buff brown (occasionally steel blue), becoming copper brown in the center when drying, strongly hygrophanous, fading to pallid white and easily bruising bluish green. Surface viscid when moist from a gelatinous pellicle, usually separable. **GILLS:** Attachment adnate to uncinate, close. Color grayish to dark cinnamon brown with the edges remaining pallid. **STEM:** 50–70 mm long by 2–3 mm thick. Equal to subequal. Pallid to brownish surface sometimes covered with fine whitish fibrils, often more yellowish toward the apex. Brittle, stuffed with loose fibers. Distinct rhizomorphs present about stem base. Partial veil thinly cobwebby, rapidly becoming inconspicuous. **MICROSCOPIC FEATURES:** Spores purplish brown, mango shaped, 10–12 by 6–7 µm. Pleurocystidia absent. Cheilocystidia 20–30(40) by 4.5–6(9) µm, fusoid with a narrow neck.

Psilocybe baeocystis is often dark brownish when moist, becoming light straw colored with bluish tones while drying out. They can be very difficult to see!

HABIT, HABITAT, AND DISTRIBUTION: Solitary to gregarious to subcespitose on decaying conifer mulch, in wood chips, or in lawns with high lignin content. Occasionally growing from fallen seed cones of Douglas fir trees. Found in the fall to early winter and rarely in the spring. (I once found it as late as June 20.) First reported from Oregon; also found in Washington, British Columbia, southwestern Alaska. Likely to be widespread along the coastal regions of the Pacific Northwest of North America. The native habitat, outside of wood chip mulch, is not yet known. I know of no reports of finding *Psilocybe baeocystis* in a natural forest. I have only found it associated with human activities, particularly in wood chips used as mulch around gardens, rose beds, and rhododendrons.

COMMENTS: Moderately active, losing significant potency in drying or from damage. Mushrooms correctly identified as *Psilocybe baeocystis* ranged from 0.15 to 0.85 percent psilocybin and up to 0.59 percent psilocin.[26,27] These early reports on potency are likely not reflective of the species' true potency, which I suspect is much higher, especially the psilocin content.

From my experience, this species is less frequently found now than decades ago. I suspect it has a close niche with Douglas fir trees. Although colonies are scarce, they can be large—sometimes abundant where wood chips and bark (especially Douglas fir) have been used for landscaping. I often find *Psilocybe baeocystis* among ivy as it encroaches over the cultivated landscapes near newly constructed buildings in Washington and Oregon. I have not yet found this species in British Columbia, although others have. This species' dark steel to gray blue color often makes it difficult to see against the background of wood chips. The dark-colored caps camouflage it until the caps lighten or whiten as they dry or are exposed to the sun, often becoming golden yellow struck through with beautiful bluish tones.

In rhododendron gardens I have frequented, *Psilocybe baeocystis*, *Psilocybe pelliculosa*, and *Psilocybe cyanofibrillosa* would often grow in proximity to each other, sometimes side by side. The incurved, sometimes enrolled cap margin and its irregular form are important features of *Psilocybe baeocystis*. This species was often misidentified as *Psilocybe cyanescens* and vice versa in the late 1960s and 1970s, leading to incorrectly identified herbarium deposits. Taxonomically, a close relative to *Psilocybe baeocystis* is *Psilocybe aztecorum*, which is synonymous with *Psilocybe quebecensis*. The mycelium is thin and wispy and difficult to detect beneath the mushrooms. In culture, the mycelium grows as a thin web, which is in stark contrast to the rich, white, and dense mycelium of other lignicolous *Psilocybe* species like *Psilocybe allenii*, *Psilocybe azurescens*, and *Psilocybe cyanescens*.

See also *Psilocybe azurescens*, *Psilocybe aztecorum*, *Psilocybe caerulescens*, *Psilocybe caerulipes*, and *Psilocybe cyanescens*.

Psilocybe caeruleoannulata SINGER ex GUZMÁN

CAP: 1–2.2 cm broad. Acutely conic to conic to convex to broadly convex, papillate, typically retaining an umbo, reddish brown to brown, hygrophanous, fading to lighter brown to yellowish to beige. Translucent-striate when moist along the cap margin, covered with a separable gelatinous pellicle. **GILLS:** Attached, ascending, sinuate to adnate, grayish violet, darkening with spore maturity to dark brown to purplish brown, with whitish edges. **STEM:** 5–45 mm long by 1–2 mm thick. Whitish to light brown, bruising bluish where injured. Adorned with a superior,

Psilocybe caeruleoannulata often has a membranous ring that is bluish toned and covered with purple brown spores. Moderately active. It bears macroscopic similarities to *Psilocybe stuntzii* and, to a lesser degree, *Psilocybe fimetaria*.

white membranous annulus, the upper side darkened from spores at maturity and often bluish. Flesh whitish, bruising bluish. **MICROSCOPIC FEATURES:** Spores 8.8–12 by 6–7.1 by 5.5–6 µm. Pleurocystidia absent. Cheilocystidia 17–20 by 5–6 µm, lageniform to vesiculose-ventricose, narrowing toward the neck, 5–8.5 by 1–2 µm.

HABIT, HABITAT, AND DISTRIBUTION: Solitary to gregarious, in pastures, on dung, in grassy and mossy manure-enriched soils. Found in southwestern Colombia around 7000 feet (2100 m) elevation and reported from Chile and Uruguay. Likely to be broadly distributed across South America.

COMMENTS: I find this species to be particularly fascinating as it has a similar appearance to the grassland forms of *Psilocybe stuntzii*. Both feature bluish superior membranous annuli. I know of no analyses for psilocybin, but I expect it would be in the range of 0.5–1 percent psilocybin with some psilocin due to the distinctive bluish of the membranous annulus and the flesh.

Psilocybe caeruleorhiza OSTUNI, ROCKEFELLER, BIRKEBAK & CANAN

CAP: 1.8–6.0 cm broad. Convex to broadly convex, hemispheric often with an incurved margin when young, expanding and becoming obtusely umbonate as the caps expand with maturity. Nutmeg colored when young, soon rusty orange to copper to orange brown or hygrophanous, fading to a lighter yellowish brown from the margin with age, bruising bluish upon handling, smooth, wrinkling with age. Margin sometimes decorated with a thin band of whitish veil remnants, soon lost with age. Context white, bruising bluish. **GILLS:** Attached, adnate to sinuate, ventricose, even, unequal, sometimes with a decurrent tooth, with three tiers of intermediate gills. Pallid when young, becoming lilac tinged to purplish brown at spore maturity. Flesh white to ochraceous. Partial veil thinly cortinate, fugacious. **STEM:** 30–120 mm long by 2–4 mm thick. Central, equal, whitish to dull white, bruising blue upon handling, surface pruinose or having fibrillose patches upward, and often tan toward

Psilocybe caeruleorhiza produces abundant whitish to bluish rhizomorphs at the bases of the stems.

Psilocybe caeruleorhiza can fruit prolifically in decorative wood mulch.

the apex, tapering toward base, sometimes peeling with age, fibrous, whitish to ochraceous with silky gloss, glabrous. Annulus absent. Stem base adorned with white radiating rhizomorphs. **MICROSCOPIC FEATURES:** Spores purple brown in deposit, subellipsoid, thick-walled, with a germ pore, 10–15 by 5.5–7.9 µm. Pleurocystidia absent. Cheilocystidia 18.2–41.9 by 6.2–13.2 µm, ventricose rostrate to lageniform to subampullaceous, lacking apical protrusion.

HABIT, HABITAT, AND DISTRIBUTION: Gregarious to clustered, collyboid, on wood mulch. Collections have been found in Ohio, Indiana, Michigan, Pennsylvania, and Iowa in November through December. Range and seasonal windows are likely to be more extensive than presently known.

COMMENTS: This new species looks strikingly like a cluster of *Agrocybe* from above until more closely examined, the differences being that *Psilocybe caeruleorhiza* has purplish brown rather than brown spores and bruises bluish.[28] See also *Psilocybe ovoideocystidiata*, but *Psilocybe caeruleorhiza* can easily be distinguished from it as *Psilocybe ovoideocystidiata* has a distinct membranous annulus, whereas *Psilocybe caeruleorhiza* does not. See also *Psilocybe caerulipes*, a much slenderer mushroom, *Psilocybe aztecorum*, and *Psilocybe serbica*. The abundant, white, thick, radiating rhizomorphs encasing wood chips are a strong indication that this species can be transplanted using stem butts or by moving lenses of rhizomorphic mycelium to newly prepared, fresh wood-chip beds). In culture on sterilized nutrient media, radiating rhizomorphs can become bluish, hence its name.

Psilocybe caerulescens MURRILL

Common Name: Derrumbe (Landslide Mushroom)

CAP: 2–9 cm broad. Obtusely campanulate to convex with a decurved margin at first, becoming convex, rarely plane in extreme age and often having either a small umbo or a slight depression in the center. Margin often bluish, translucent-striate halfway to the center portion of the cap and hanging with fragile whitish veil remnants (appendiculate). Deep olive brown to very dark brown in young specimens, becoming dark brown, sometimes with blackish tones, strongly hygrophanous, fading with age to a dark reddish brown to chestnut brown near the disc and remaining darker and translucent-striate toward the margins when still moist. Margin incurved and often inrolled when young, expanding and even when mature. Surface smooth and slightly viscid to lubricous when moist, pellicle thinly gelatinous, sometimes not separable. Flesh whitish to dingy brown, moderately thick, and bruising bluish green. **GILLS:** Attachment sinuate to adnate, close and broad. Color grayish to sooty brown with the edges remaining whitish. **STEM:** 40–120 mm long by 2–10 mm thick. Mostly equal but often narrowing into a long pseudorhiza. Covered

Psilocybe caerulescens grows in the southeastern United States, southern Mexico, and south to northwestern South America. It is often associated with landslides.

at first with an impermanent whitish layer of fibrils, overlaying a more sordid brown smooth surface underneath. Upper regions of the stem characteristically adorned with whitish fibrillose patches. Partial veil cortinate, whitish and copious at first, but soon disappearing. Flesh stuffed and fibrous; bruising bluish green, whitish rhizomorphs (bluish when disturbed) present about the base of the stem. **MICROSCOPIC FEATURES:** Spores dark purplish brown, subrhomboid to subellipsoid, 6–8 by 4–6 µm. Basidia 4-spored, occasionally 2-spored. Pleurocystidia scattered to absent. Cheilocystidia 15–22 by 4.5–6 µm, fusoid with a flexuous neck 1–2.5 µm broad.

HABIT, HABITAT, AND DISTRIBUTION: Gregarious to cespitose, rarely solitary, in the late spring and summer on disturbed or cultivated grounds often devoid of herbaceous plants. Preferring muddy orangish brown soils. First reported from near Montgomery, Alabama, by Franklin Sumner Earle in 1923 on sugar cane mulch.[29] Recently, this species has been found in western Louisiana to as far north as North Carolina. It is likely more widely distributed. *Psilocybe caerulescens* is widespread throughout the central regions of Mexico and found as far south as Venezuela.

COMMENTS: A potent species; a very high dose (approximately thirteen pairs of this mushroom) was ingested by R. Gordon Wasson during his inaugural session with María Sabina, the renowned Mazatec healer. Heim and Wasson reported 0.20 percent psilocybin and 0 percent psilocin from Albert Hofmann.[30] This analysis vastly understates the amount of psilocybin

Psilocybe caerulescens can display a wide range of forms as it matures. The caps and gills darken as spore production increases. Note spore cast on top of mushroom caps, upper right.

and psilocin from freshly dried specimens. See also *Psilocybe weilii*, to which this species is very closely related. *Psilocybe weilii* has abundant pleurocystidia, whereas *Psilocybe caerulescens* was originally reported not to. However, a reexamination by Alan Rockefeller detected some pleurocystidia in the type collections of *Psilocybe caerulescens*. Nevertheless, Bradshaw et al. found differences in the DNA analysis between these two taxa.[31] They may or may not remain separate species upon further examinations. In any event, they are very closely related.

Psilocybe caerulipes (PECK) SACCARDO

Common Name: Blue Foot

CAP: 0.5–5 cm broad. Bluntly conic becoming conic-campanulate to broadly convex to plane with age and can be slightly umbonate. Margin incurved at first, sometimes tinged greenish blue, irregular, closely translucent-striate when moist, and decorated at first with fibrillose veil remnants. Cinnamon brown to dingy brown, hygrophanous, fading to pale ochraceous buff. Surface viscid when moist from a gelatinous pellicle, but soon becoming dry and shiny. Flesh thin, pliant, and bluing where bruised. **GILLS:** Attachment adnate to sinuate to uncinate, close to crowded, narrow, with edges remaining whitish. Color sordid brown at first becoming rusty cinnamon. **STEM:** 20–60 mm long by 2–4 mm thick. Equal to slightly enlarged toward the base. White to buff at first, with the lower regions dingy brown at maturity, bluing where bruised. Surface powdered toward the apex and covered with whitish to grayish fibrils downward. Flesh stuffed with a pith and solid at first, soon becoming tubular. Partial veil thinly cortinate and forming an evanescent fibrillose annular

Psilocybe caerulipes, the Blue Foot, is distributed across much of northeastern North America.

zone in the superior region of the stem, if at all. **MICROSCOPIC FEATURES:** Spores dark purplish brown in deposit, ellipsoid, 7–10 by 4–5 µm from 4-spored basidia. Spores from 2-spored basidia are larger. Pleurocystidia absent. Cheilocystidia 18–35 by 4.5–7.5 µm, lageniform, with a thin neck, sometimes forked, 1–2.5 µm broad at apices.

HABIT, HABITAT, AND DISTRIBUTION: Solitary to clustered on hardwood slash and debris and on or about decaying hardwood logs, particularly birch, beech, and maple, especially along river systems. Growing in the summer to fall after warm rains. Widely distributed east of the Great Plains throughout the Midwest and eastern United States, and in beech (*Fagus* spp.) forests in Mexico, the state of Hidalgo, and the southernmost range of beech in North America.

COMMENTS: Moderately active; no analyses published. The bluing reaction is variable, more evident in drying. Although widely distributed, *Psilocybe caerulipes* is not as frequently found as *Psilocybe ovoideocystidiata*, although they can co-occur in the same habitat, typically separated by the seasons. *Psilocybe caerulipes* thrives in the fall, whereas *Psilocybe ovoideocystidiata* prefers the spring. Recent molecular analysis shows that *Psilocybe caerulipes* is one of the more ancestral species, estimated to have appeared around 25 million years ago.[32] See also *Psilocybe aztecorum*, *Psilocybe cyanescens*, and *Psilocybe ovoideocystidiata*.

Psilocybe callosa (FRIES ex FRIES) QUELET

CAP: 0.5–3.0 cm broad. Conic at first, expanding to convex, campanulate, and eventually broadly convex, and typically not sharply umbonate. Surface smooth, translucent-striate near the margin, viscid when moist from a separable gelatinous pellicle. Dark grayish brown to cinnamon brown, fading to straw or light yellow in drying. Flesh sometimes bruising bluish when injured, but not consistently. **GILLS:** Attachment adnate, sometimes subdecurrent, and tearing free from the stem in drying. Chocolate brown when mature with whitish edges. **STEM:** 40–70(130) mm long by 2–3 mm thick. White to yellow to yellowish brown. Equal, straight to

Psilocybe callosa is a grassland species, also thought by some to be synonymous with *Psilocybe strictipes*. This species typically does not bruise bluish, lacking psilocin, and is estimated to contain modest levels of psilocybin, but much less than the similar *Psilocybe semilanceata* (Liberty Cap). *Psilocybe callosa* lacks the sharp nipple (umbo) that is distinctive of *Psilocybe semilanceata*. *Psilocybe callosa* typically grows gregariously, but scattered, rarely in clusters.

Psilocybe callosa is strongly hygrophanous, soon fading in the sunlit grasslands in which it frequents.

Psilocybe callosa has a gelatinous pellicle—separable only when wet—which easily distinguishes it from many other little brown non-*Psilocybe* mushrooms that grow in grasslands.

flexuous, typically tough, cartilaginous, decorated with fibrillose patches from veil remnants and adorned with basal mycelium that can bruise bluish. Partial veil cobwebby, thin, fragile, and rarely leaving an annular zone on the upper regions of the stems. **MICROSCOPIC FEATURES:** Spores dark purple brown in deposit, subellipsoid to suboblong, 10–12 by 5.5–8 µm. Pleurocystidia absent. Cheilocystidia 21–45 by 7–10 µm, lageniform with an extended neck, 2–3.5 µm thick.

HABIT, HABITAT, AND DISTRIBUTION: Fruiting in the late summer to fall in the Pacific Northwest of North America, England, northern Europe (France, Holland, Sweden, Germany, Czechoslovakia), Siberia, Chile, and reportedly South Africa. Typically found in rich, grassy areas such as lawns, along roadsides, and in fields. Typically not found on dung, although common in fields with and without manure.

COMMENTS: *Psilocybe callosa* is a lesser-known species to most collectors of psilocybin mushrooms. One of the reasons, I suspect, is its absence of notable features. This species is difficult for most experts to recognize. It does not bruise bluish, lacks an umbo, and looks nondescript compared to other species like the usually nippled *Psilocybe semilanceata* with which it can co-occur. *Psilocybe callosa* and *Psilocybe semilanceata* are closely related to the point that they may be two divergent forms within the same generalized species group.

Another closely related mushroom name that has been applied here is *Psilocybe strictipes*, a mysterious, if not mythical, species whose provenance in taxonomy remains complicated by a lack of recent collections; some herbarium samples may be mixed collections with *Psilocybe semilanceata*. Gastón Guzmán attempted to clear up the confusion surrounding *Psilocybe callosa* by suggesting it is conspecific with *Psilocybe strictipes*, which was first published as a conifer-loving mushroom by Singer and Smith, but DNA analysis suggests *Psilocybe strictipes* belongs to *Phaeonematoloma*.[33,34] I am also doubtful that *Psilocybe strictipes* is a valid species. There is considerable confusion between these taxa that needs sorting out.

For most mushroom hunters, *Psilocybe callosa* is similar in its morphology to the pastoral *Psilocybe semilanceata* and less so to *Psilocybe liniformans*. The lack of a sharp umbo is the only macroscopic feature I know that can delineate this species from its closest ally, *Psilocybe semilanceata*, with which it is often confused. Microscopically, *Psilocybe callosa* has smaller and narrower spores than those of *Psilocybe semilanceata* but otherwise is very similar and may, in fact, be two varieties within the same species. Also, stem length is not a good delineator as it is usually influenced by the height of the grass clumps from which it grows. The stem base of *Psilocybe callosa* is typically tightly attached to dead grass thatch. The name *callosa* refers to the tough or hardened texture of the mushroom when dried, especially the stem.

One variety of *Psilocybe callosa* grows abundantly in west-central Oregon's Willamette Valley in the corridor from Portland to Eugene, in close association with highland bent grass (*Agrostis tenuis*), where thousands of acres are dedicated to the commercial cultivation of grass seed—a major industry of that region. The prolific fruitings of *Psilocybe callosa* in these grasslands and the subsequent distribution of spore-dusted seeds represents a huge launching platform to inoculate far-away lawns and, for instance, golf courses. The potential distribution of this species through the commercial distribution of lawn seed is mind-boggling. It is likely to be much more common than presently realized. Please be sure the mushrooms you pick have purple brown spores, as many rusty brown–spored mushrooms like species of *Galerina* and *Pholiotina* co-inhabit the same ecosystem.

Psilocybe cubensis (EARLE) SINGER = *Stropharia cubensis* EARLE

Common Names: Golden Tops, Golden Teachers, Cubies, San Isidro

CAP: 1.5–8.0 cm broad. Conic campanulate, often with an umbo at first, becoming convex to broadly convex and finally plane in age with or without an obtuse or acute umbo. Reddish cinnamon brown in young fruitbodies, becoming lighter with age to more golden brown, fading to pale yellow or white near the margin with the umbo or the center region remaining darker and cinnamon brown. Surface viscid to smooth when moist but soon dry; universal veil leaving fragile, spotted whitish remnants on cap but soon becoming smooth overall. Flesh whitish, soon bruising bluish. **GILLS:** Attachment adnate to adnexed, soon seceding, close, narrow to slightly enlarged in the center. Pallid to grayish in young fruitbodies, becoming deep purplish gray to nearly black in maturity, often mottled.

Psilocybe cubensis, the Golden Top, is a regal species that displays many forms. Its rich golden color, thick membranous partial veil, and ring (adorned with veil remnants on the caps) are soon washed away by rain or wind, or fade with exposure to sun. Note bluish reactions at bases of stems. This iconic species appears to be at the center of a constellation of species sharing the same overall morphology. See also *Psilocybe ochraceocentrata* on page 178.

The cow-dung-loving *Psilocybe cubensis* has gills that darken with spore maturity. The spore cast accumulates on the upper surface of the membranous annulus. Note bluing on stems from handling.

STEM: 40–150 mm long by 5–15 mm thick. Thickening and often curved toward the base. Whitish overall but may discolor to yellowish; bruising bluish green where injured. Surface smooth to striated at the apex and dry. Partial veil typically thickly membranous, frequently bluish toned at the veil–cap connection, dropping to leave a well-developed white, persistent membranous annulus that often bruises bluish and soon becomes colored with purplish brown spores. **MICROSCOPIC FEATURES:** Spores dark purplish brown to violaceous brown, subellipsoid, double walled, 11.5–17 by 8–11 µm. Basidia 2- to 3-spored, but usually 4-spored. Pleurocystidia subpyriform, sometimes mucronate, swollen at the apex, 18–30 by 6–13 µm. Cheilocystidia fusoid ventricose with an obtuse or subcapitate apex, sometimes sublageniform, 17–32 by 6–10 µm with the narrow necks 3–5 µm broad.

HABIT, HABITAT, AND DISTRIBUTION: Scattered to gregarious on cow dung or well-manured grounds in the spring, summer, and fall. This species is likely to be associated with the dung of many ungulates (hoofed mammals). Not yet reported from buffalo or bison dung. Mycologist Gary Lincoff found this species group on elephant dung in India, suggesting that *Psilocybe cubensis* has a broader range than just ungulates, perhaps tracing its lineage to many megafauna herbivores, some predating the appearance of cattle.

Psilocybe cubensis is circumpolar, widely growing in the tropical and subtropical regions of the world. Found throughout the southeastern United States, Mexico, Cuba, Central America, northern South America, and the Far East, including India, Thailand, Vietnam, and Cambodia, and regions of Australia (Queensland). Surprisingly, collections by academic mycologists in Africa are lacking, although a similar, newly named species, *Psilocybe*

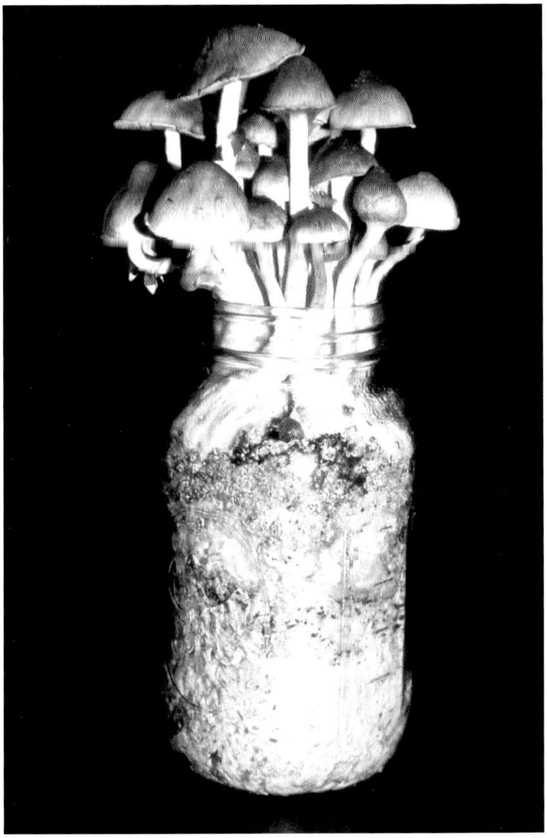

Psilocybe cubensis can easily be grown on a wide variety of sterilized grain. Here is a fruiting of mine on rye grain. Rice, wheat, millet, sunflower, sorghum, corn, and many other grains support fruiting, although some prefer rice or a rice/millet/sunflower combination. Commercial bird seed formulations are also commonly used.

ochraceocentrata, is known, albeit thus far, only from southern Africa. These species are closely interrelated as is *Psilocybe aquamarina*. *Psilocybe jaliscana* is synonymous with *Psilocybe cubensis*.

Typically, the most prolific fruitings of this species are seen in the 2 months prior to the hottest period of the year. During extremely hot temperatures, this species is less frequently encountered; its upper limits seem to be around 90°F (32°C) ground temperature. In the southeastern United States, May and June are the best months for picking, although it can be found until January. A highly adaptive species, *Psilocybe cubensis* has spread throughout the warmer, moister regions of the world and is strongly associated with egrets. These birds often have a symbiotic relationship with cattle, removing biting insects like fleas, flies, and ticks. Egrets also take advantage of eating the grasshoppers that alight as cattle walk, bringing them in close contact with sporulating fruitbodies of this mushroom species. Egrets can also migrate across continents. Many bird species are known to spread mushroom spores. This cattle–insect–bird–mushroom bridge is another ingenious way mushrooms have evolved to increase their survival by using birds to spread their spores.

Now, with its widespread cultivation, *Psilocybe cubensis* is well poised to migrate further, even into temperate regions. The furthest north in North America that I know this mushroom has naturalized is the state of Arkansas. With warming climates, it is likely to spread northward, as with many species.

COMMENTS: A majestic species, most experts agree that *Psilocybe cubensis* was likely introduced to the Americas when Spanish colonists and missionaries first brought cattle with them in the late 1400s and 1500s, and then repeatedly thereafter. Interestingly, we do not yet know the origin of this species in Africa. This mushroom came to be known by locals in Mesoamerica as San Isidro, the Catholic patron saint of farmers. Initially disdained by Indigenous peoples in favor of native psilocybin species, *Psilocybe cubensis* is increasingly being used in Indigenous ceremonies because it can be sustainably grown indoors and is available year-round.

Today, *Psilocybe cubensis* is the most widely used and cultivated psilocybin mushroom in the world. It can contain up to 5 percent psilocybin

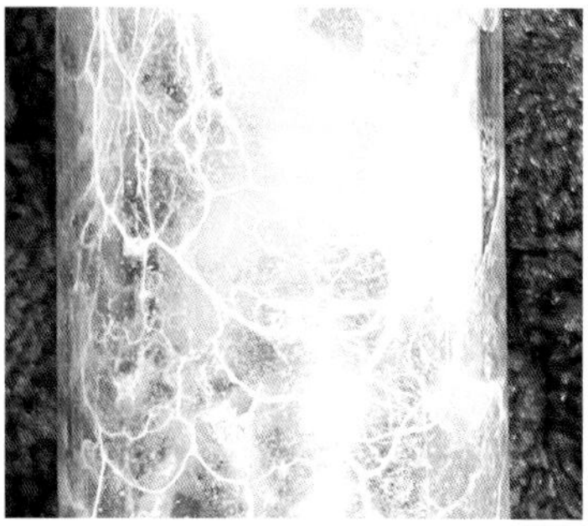

Psilocybe cubensis culture in a test tube, showing classic rhizomorphs.

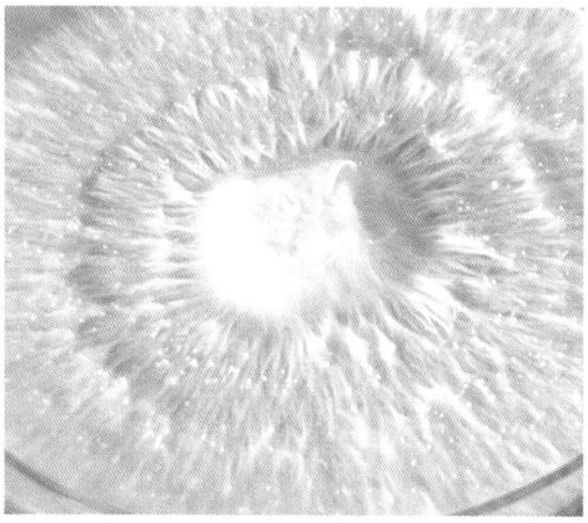

Psilocybe cubensis forming primordia in a nutrient-filled Petri dish.

Psilocybe cubensis mushrooms maturing on a nutrient-filled Petri dish. Note bluing reactions and spore cast.

and psilocin by dry weight. However, a recent analysis of cultivated strains detected 0.85–1.45 percent combined psilocybin and psilocin.[35]

I would not be surprised if some phenotypes express more than 5 percent combined psilocybin and psilocin. Air-dried specimens can lose 20 to 40 percent of their potency in a year's time. Numerous factors affect loss in potency, including strain type, substrates, and drying and storage conditions. Most *Psilocybe cubensis* mushrooms have around 1 to 2 percent psilocybin/psilocin, or 10 to 20 mg psilocybin/psilocin per dried gram of mushrooms. Curiously, I have found that when this species is stored in a cool, dry place ranging from 40 to 65°F (4 to 18°C), the amount of psilocybin was nearly 2 percent after 2 years. Cool, dry storage in vacuum-packed food storage bags may be better than freezing or freeze drying. My experience matches a study by Gotvaldová et al., which also surprisingly found that *Psilocybe cubensis* stored at room temperature had much higher potency than deep frozen (-112°F, -80°C) or freeze-dried specimens.[36] Not surprisingly, strains also differ substantially in their psilocybin content independent of substrate or storage conditions. There are even strains of *Psilocybe cubensis* that, upon analysis of fresh fruitbodies, had no detectable psilocybin, although this is extraordinarily unusual. As with all psilocybin mushrooms, please be careful and err on the side of caution when ingesting, so you are confident of having a maximally predictable dose. Potency is a consequence of many factors that include genetics, substrate, mushroom mass, drying method, light and oxygen exposure, temperatures, and lastly duration of storage.

Compared to most other psilocybin species, *Psilocybe cubensis* is the easiest to cultivate indoors. The mycelium is aggressive, produces characteristic white, rhizomorphic cultures, and fruits on a wide array of substrates. That it fruits abundantly on grain (rice, rye, wheat, corn, millet, sorghum, oats, and sunflower seeds) positions this species with unique advantages for end-users who can easily grow it at home and who lack the skills or space for composting. Using pressure cookers, some elementary lab equipment, and following techniques described in good cultivation guides (including *The Mushroom Cultivator*[37] or *Growing Gourmet and Medicinal Mushrooms*[38]), cultivated *Psilocybe cubensis* allows reliable access year-round. Cultivation is inexpensive and intimately bonds the cultivator with the sacrament at hand. Additionally, the number of strains sourced from the wild, isolated, and tested is resulting in a bizarre, diverse array of phenotypes, from short fat-stemmed spheroid strains to giant "golden teachers." (Note that these common names are often used for marketing purposes and do not reflect one strain.) As thousands of growers continue to innovate, I am excited to see what new strains and techniques arise from the experiments of citizen psilocybin mushroom scientists. Not unsurprisingly, the wild strain diversity of this species is far greater than those currently under cultivation.[39]

Some people are allergic to *Psilocybe cubensis* spores. In one study of asthmatic children, 100 percent of them reacted negatively to spores of this species, highlighting its potential as an allergen.[40] Spores, which do not contain psilocybin, emerge from flesh that does. Arguably, the more spores, the less potent the mushrooms have become. Additionally, once mushrooms sporulate, they tend to rot quickly, sometimes from airborne microorganisms or bacteria brought in by flies attracted to spore release. For at least these three reasons, harvesting *Psilocybe cubensis* when the partial veils are still intact is ideal (see photo on page 64). Ingesting younger mushrooms is better than old ones, as with many species.

See also *Psilocybe ochraceocentrata*, which is macroscopically very similar to *Psilocybe cubensis* and a close relative, first described from specimens collected in South Africa.

Psilocybe cyanescens WAKEFIELD

Common Names: Cyans, Wavy Caps

CAP: 1–4(7) cm broad. Bluntly conic to conic-convex at first, with incurved irregular margin, soon expanding to broadly convex to nearly plane in age with a pronounced undulating or wavy margin, flaring in maturity, that is typically translucent-striate when wet. Young specimens can be dark chestnut brown, becoming more caramel colored with age, hygrophanous, differentially yellowing as they dry. Surface smooth and viscid when moist from a sometimes-separable gelatinous pellicle. Context nearly concolorous with the cap except for the region where stem attaches to cap, where a white pithy mycelium is evident. The cap bruises bluish green when damaged, in age, and from freezing and thawing. **GILLS:** Attachment adnate to subdecurrent, close to subdistant, broad, with two tiers of intermediate gills. Color light brown, darkening to cinnamon brown becoming dark smoky brown with the edges paler. **STEM:** 20–110 mm long by 2.0–7.0 mm thick. Often curved and enlarged at the base and tough. Whitish overall, readily

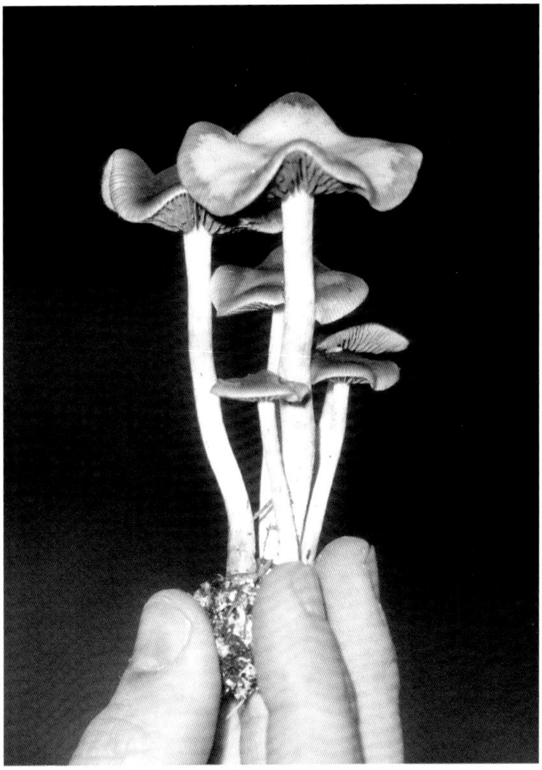

Classic phenotype of *Psilocybe cyanescens*, the Wavy Cap. Has a distinctive undulating cap margin.

Psilocybe cyanescens often fruits in clusters and can have short stems in exposed wood chips.

bruising bluish. Surface silky, covered with fine fibrils and often with long whitish rhizomorphs emanating from the stem base, which tends to be darker in color. Partial veil copiously cobwebby, white, stretching, breaking, and rapidly deteriorating to an obscure annular zone that soon disappears. Purple brown spores often decorate the scattered fibrils on the stem. Annular zones typically absent. **MICROSCOPIC FEATURES:** Spores dark purplish brown in deposit, elongate-ellipsoid, 9–12 by 5–8 µm. Although few pleurocystidia were reported by Singer and Smith's examination of the 1946 collection that Wakefield deposited at the Kew Gardens near London, the specimens I have collected from the Pacific Northwest of North America have abundant pleurocystidia, 17–33 by 5.0–8.8 µm, fusoid ventricose to subpyriform, mucronate, more populous near the gill edge. Cheilocystidia (12)16–27(30) by (5)6.6–8.8 µm, sublageniform to fusoid-ventricose, cylindrical at base, with an extended single or split neck 6 by 1.5–3.5 µm.

HABIT, HABITAT, AND DISTRIBUTION: Scattered to gregarious in woody debris, among leaves and twigs, in wood chips or sawdust, or in debris fields rich with decomposing deciduous wood. This species loves gardens and is often found wherever wood mulch is laid down: along edges of lawns bordered with wood chips, wood-chip paths, and in heavily mulched rhododendron and rose gardens. I think this species may be an endophyte, living within alder (*Alnus*) trees. I have never found this species in a natural woodland but have seen it spring voluntarily from wild alder trees that have been chipped and used for landscaping. Found in the fall to early winter. Reported from the western coastal regions of North America (from San Francisco to southern Alaska) and now also widespread throughout the United Kingdom and across much of temperate Europe, from Italy to Germany to Spain to Norway. This species is moving around the world!

COMMENTS: Potent. Our collection of this species, reported by Beug and Bigwood,[41] showed 1.68 percent psilocybin and 0.28 percent psilocin. Jochen Gartz [42] found 0.30 percent psilocybin, 0.51 percent psilocin, and 0.02 percent baeocystin. Stijve and Kuyper[43] reported 0.85 percent psilocybin, 0.36 percent psilocin, and 0.03 percent baeocystin. Bradshaw et al. found psilocybin from 0.3 to 1.56 percent and psilocin from 0.02 to 0.52 percent. A recent analysis I did showed nearly 3 percent psilocybin and psilocin in 2-year-old specimens stored at room temperature. Surprisingly, *Psilocybe cyanescens*, stored in an herbarium for nearly 50 years, still had many micrograms/milligram potency, which is remarkable because psilocin is unstable compared to the much more stable psilocybin. Other specimens less than 20 years of age had no detectable psilocybin or psilocin. This wide range of variability underscores that many factors influence potency. None of the above analyses discerned between the age of the fruitbodies. Besides the phenotype and growth conditions, I think the maturity of the fruitbodies tested has a major influence. Mushrooms harvested when they are young, before heavy sporulation, prior to bluing, will likely have higher potency than heavily sporulating, more mature specimens. Air drying in front of a fan, at high flow but with no heat, is my preferred method for drying this (and many) species.

Psilocybe cyanescens can have short stems when growing from exposed wood chips.

Many mycologists suspect that *Psilocybe cyanescens* likely colonized Europe through the collections of wild and domesticated plants, such as rhododendrons sent from the Pacific Northwest to the Kew Gardens, a major nexus for the importation of foreign botanicals. Botanical gardens likely have become hubs for spreading many nonnative species. In fact, Singer and Smith noted in their monograph that "It is not impossible that this species was introduced to Europe with exotic plant material."[44]

Interestingly, slugs seem to have a distaste for this potent psilocybin species. Even when slugs are placed directly on top of *Psilocybe cyanescens*, they soon depart. However, slugs placed on Garden Giant (*Stropharia rugoso-annulata*) devour it. When I did a choice test by placing ten slugs in a box and these two species on opposite sides, the slugs migrated to the Garden Giant. This observation supports the hypothesis that psilocybin could have evolved to prevent being parasitized by slugs and snails.

Psilocybe cyanofibrillosa GUZMÁN & STAMETS

Common Names: Rhododendron Psilocybe, Blue-Haired Psilocybe

CAP: 1.4–3.5 cm broad. Conic to convex to broadly convex, eventually plane in age. Color deep chestnut brown to caramel brown, hygrophanous, fading in drying to pale tan to yellowish brown, even dingy grayish white. Surface viscid when moist from a separable gelatinous pellicle. **GILLS:** Attachment adnate to adnexed to slightly subdecurrent in age, light grayish when young, becoming purplish brown with maturity with whitish edges. **STEM:** 30–70 mm long

Psilocybe cyanofibrillosa caps are even along the edges, especially when young.

Psilocybe cyanofibrillosa is so named for the bluish-toned hairs that can be seen on the stem.

by 2–4 mm thick, straight to flexuous, equal to enlarged near the base, longitudinally striate, and adorned with fine fibrils that become bluish when handled. Yellow brown to light tan underneath. Partial veil white, cortinate, copious when young, thinning in age, leaving fragile fibrillose veil remnants as annular zone on the upper regions. Flesh brownish, bruising bluish. **MICROSCOPIC FEATURES:** Spores purplish brown in deposit, subellipsoid, 9–12 by 5.5–7 µm. Basidia 4-spored, rarely 2-spored. Pleurocystidia absent. Cheilocystidia fusiform to lanceolate, 22–33 by 5.5–7 µm, with an elongated neck, 1–1.5 µm at apex.

HABIT, HABITAT, AND DISTRIBUTION: Growing gregariously to scattered and primarily a coastal species found from California (Eureka/Arcata) north to British Columbia, Canada. Associated with bush lupines and especially common on flood plains on river estuaries flowing into the Pacific Ocean. Also frequently found in coastal rhododendron gardens and nurseries or where conifer wood chips have been used for landscaping. A nonprofit foundation devoted to rhododendron species, funded by a well-known lumber company, created a 20-plus-acre rhododendron park mulched with massive amounts of conifer wood chips, providing an ideal habitat for this species; *Psilocybe cyanescens* and *Psilocybe baeocystis* also thrived there for many years.

COMMENTS: Mildly active and likely containing more than 0.05 mg per gram of psilocybin and 1.4 mg per gram of psilocin as initially reported by Stamets et al.[45] Although this species has a fairly strong bluing reaction due to its comparatively high psilocin content, it loses a lot of potency from handling and in drying. *Psilocybe cyanofibrillosa* has only been reported from the Pacific Coast region of North America. See *Psilocybe pelliculosa*, one of its closest relatives. Additionally, see *Psilocybe allenii, Psilocybe azurescens*, and *Psilocybe cyanescens*, which have abundant pleurocystidia, whereas *Psilocybe cyanofibrillosa* does not. If wood mulch contains a mixture of hardwoods and conifers, these species can co-occur. I have only found *Psilocybe cyanofibrillosa* in conifer sawdust, suggesting that it might be an endophyte, latent within living Douglas fir trees, fruiting postmortem.

Psilocybe fimetaria can grow gregariously on horse dung.

Psilocybe fimetaria (P.D. ORTON) WATLING

CAP: 0.5–3.6 cm broad. Conic to convex, eventually subcampanulate, expanding to broadly convex, often with an obtuse umbo. Surface smooth to translucent-striate near the margin, viscid when moist from a thick, separable gelatinous pellicle. Color pale reddish brown to honey to ochraceous, hygrophanous, fading in drying to ochraceous buff to yellowish olive to yellowish. Flesh whitish to honey colored, inconsistently bruising bluish where injured. **GILLS:** Attachment adnate, sometimes uncinate, whitish clay at first, eventually dark purplish brown at maturity, with whitish edges. **STEM:** 20–90 mm long by 1–4 mm thick. Equal to slightly swollen at the base. Color whitish at first, soon reddish brown or honey colored, and sometimes with grayish to bluish green tones. Surface covered with whitish fibrillose patches to intermittent superior densely fibrillose zones that develop from a thickly cortinate partial veil evident in younger specimens. **MICROSCOPIC FEATURES:** Spores dark purplish brown, subellipsoid or ellipsoid, 9.5–16 by 6.5–9.5 µm. Pleurocystidia absent. Cheilocystidia 20–32 by 4–8 µm, ventricose-fusiform or lageniform with a narrow neck, often flexuous, 4–15 by 0.5–1.5 µm, occasionally branched.

Psilocybe fimetaria is a classic dung-dwelling, psilocybin-active Psilocybe but does not typically bruise bluish. When it does bruise bluish, the bluish tones are most evident in the mycelium at the base of the stem.

Psilocybe fimetaria can have hemispheric to conic caps, sometimes with a blunt umbo. Note the purple brown spore cast on the fibrillose remnants of the partial veil on the upper regions of the stems.

HABIT, HABITAT, AND DISTRIBUTION: Growing singly to gregariously on horse manure, in grassy areas or rich soils, and often fruiting in large rings. Known from Great Britain and Europe (Finland, Norway, Czech Republic). Collections have also been reported from Canada (British Columbia, New Brunswick) and the northwestern United States (Washington, Oregon, Idaho). Generally found in October and November. In Chile, this mushroom has been collected in August. Its fruiting range is probably much greater than presently reported.

COMMENTS: Moderately active to inactive.[46] At first glance, this species shares a strong resemblance to hemispherically capped *Deconica coprophila* and *Psilocybe liniformans*, which has a peelable gill edge (see page 166). *Psilocybe fimetaria* can sometimes be conic capped, but usually not as sharply conic as is typical with *Psilocybe semilanceata*. *Psilocybe semilanceata* typically prefers cow and sheep pastures and does not grow from horse dung, which is the preferred dung for *Psilocybe fimetaria*. Both can have superior fibrillose annular zones dusted with purplish brown spores as the mushrooms mature. *Psilocybe hispanica* may be synonymous with *Psilocybe fimetaria*. *Psilocybe subfimetaria* may just be a varietal of *Psilocybe fimetaria* as might be *Psilocybe hispanica*, which can likewise grow on horse dung.

Psilocybe hispanica GUZMÁN

CAP: 0.25–1 cm broad. Conic to convex, sometimes with a slight umbo. Brown to hazelnut, smooth, hygrophanous, fading upon drying to yellowish brown. **GILLS:** Attached, adnate, violaceus brown with spore maturity, with whitish edges. **STEM:** 16–25 mm long by 0.5–1 mm thick, even, and slightly enlarging at the base. Flesh bruising greenish blue. **MICROSCOPIC FEATURES:** Spores dark purplish brown, ellipsoid, 12–14.5 by 6.5–8 μm. Pleurocystidia absent. Cheilocystidia present.

HABIT, HABITAT, AND DISTRIBUTION: Growing in horse dung in grasslands at elevations of 3600–7500 feet (1700–2300 m) in the Pyrenees Mountain range in northern Spain and southwestern France. This species is likely more widely distributed than presently reported.

COMMENTS: This species was first described by Gastón Guzmán in 2000[47] and again by Roberto Fernández-Sasia in 2006.[48] In 2011, Brian Akers et al. proposed that the roughly 6000-year-old cave art at Selva Pascuala in the Iberian Mediterranean Basin of Spain features classic psilocybin mushrooms, suggesting that *Psilocybe hispanica* or another psilocybin mushroom was used ritualistically (see page 27).[49] However, the

Psilocybe hispanica grows in the highlands of Spain and is very similar to, if not conspecific with, *Psilocybe fimetaria*.

Psilocybe hispanica fruits on horse dung.

art depicts long-horned cattle with the mushrooms underneath, whereas *Psilocybe hispanica* prefers horse dung. Having collected *Psilocybe semilanceata* many times, I have often found it in pastures where horses and cows comingle. In either case, the cuspidate shape of the mushrooms depicted in the cave art are *Psilocybe*-like, in contrast to the oft depicted hemispherical shaped, nonumbonate, and ringed *Agaricus* (portobello) or *Amanita* forms characteristic of many non-*Psilocybe* mushrooms. Interpreting art without mycological specimens presents hurdles for any conclusions, but the circumstances here are worthy of consideration.

Deconica coprophila (formerly *Psilocybe coprophila*) can co-occur with *Psilocybe hispanica* but does not bruise bluish and is psilocybin-inactive. *Psilocybe hispanica* is particularly fascinating to me as it is one of three *Psilocybe* species that is known to grow in horse dung, can bruise bluish, and occurs in temperate regions of Europe. Another species found on horse manure that is bluing and psilocybin-active include *Psilocybe fimetaria*, to which *Psilocybe hispanica* is closely related. The similarity of these two species merits further investigation. They may be, in fact, the same species, which would mean the taxon *Psilocybe fimetaria* would take precedence, and *Psilocybe hispanica* would be taxonomically subjugated as a synonym. The psilocybin-active *Psilocybe liniformans* and *Panaeolus cinctulus* can also grow concurrently in the same habitat: horse dung paddies or in grasses enriched with decomposing horse dung.

Psilocybe hoogshagenii HEIM *sensu lato*

Common Names: Pajaritos de Monte (Little Birds of the Woods)

CAP: 0.7–3 cm broad. Conic to campanulate to convex with an acute, extended nipple (up to 4 mm long). Surface slightly viscid when wet, smooth, often ridged halfway to the disc. Reddish brown to orangish brown to yellowish, hygrophanous, fading in drying to straw colored, and bruising blue or blue black. **GILLS:** Attachment adnate to adnexed, pale brown to coffee colored, eventually purplish black at maturity. **STEM:** 30–110 mm long by 1–3 mm thick. Equal to slightly thickened near the base, flexuous, sometimes twisted. Whitish to brownish red near the base, easily bruising bluish to bluish black. Partial veil thinly cobwebby, fragile, soon disappearing. **MICROSCOPIC FEATURES:** Spores dark purplish brown in deposit, rhomboid to subrhomboid, 5–9.6 by 4–5.6 µm. Basidia 4-spored, rarely 2-spored. Pleurocystidia 16–36 by 8–12 µm, ventricose to clavate, often irregular. Cheilocystidia (15)19–35 by 4.4–6.6 µm, lageniform, narrowing into a long neck, 1–3 µm, either acute or subcapitate at the apex.

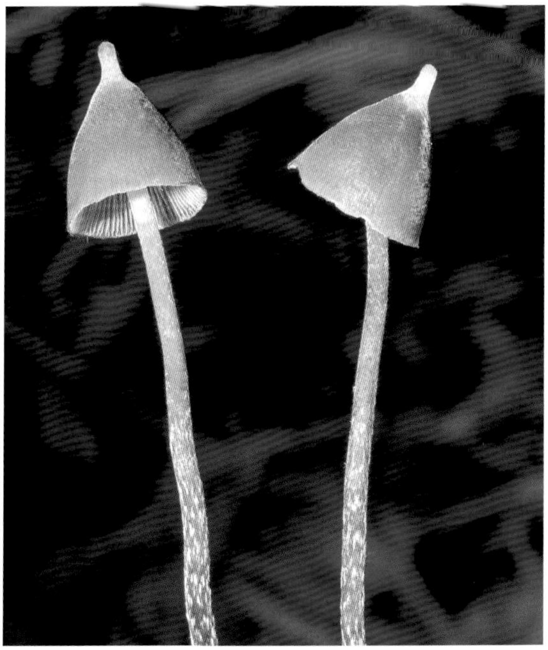

Psilocybe hoogshagenii is a potent psilocybin species distinguished by a very pronounced nipple. This species frequents the debris fields created when maintaining coffee plantations and forests.

HABIT, HABITAT, AND DISTRIBUTION: Singly to gregariously in muddy clay soils in subtropical coffee plantations. Found March through November in Mexico (Puebla, Oaxaca, Chiapas), Costa Rica, and in February in Argentina. In Costa Rica, it is common along trails 4700 to 6200 feet (1440 to 1900 m) in elevation, threading through oak and cedar forests, and within coffee plantations. *Psilocybe hoogshagenii* is probably much more widely distributed than currently reported.

COMMENTS: One of the most unusual *Psilocybe* species discovered, this mushroom is quite potent despite the modest levels reported by Heim and Hofmann, who found 0.6 percent psilocybin and 0.1 percent psilocin from aged specimens. I am struck that *Psilocybe hoogshagenii, Psilocybe stametsii,* and a yet-unnamed species (see page 227) all have "dancing nipples." I am confounded as to why this feature would be evolutionarily useful. Would insects prefer this perch? Is this just a random artifact of evolution?

Gastón Guzmán (1984) reported that this mushroom grows between 3300 and 5900 feet (1000 and 1800 m) in elevation.[50] *Psilocybe hoogshagenii* is commonly seen by coffee growers, who report massive flushes coming up in unison and then disappearing.

Psilocybe hopii GUZMÁN, J. GREEN

CAP: 1–3 cm broad. Convex to slightly umbonate, expanding in age with an undulating margin, viscid when moist. Surface smooth. Covered with a thick gelatinous pellicle, intermittently separable. Translucent-striate along the margins, which sometimes can have thin, clinging veil remnants. Dark caramel when young, becoming yellowish brown, hygrophanous, discoloring to yellowish or whitish. **GILLS:** Ascending, sinuate, yellowish brown to purplish brown with spore maturity, edges fringed whitish. **STEM:** 40–100 mm long by 2–5 mm thick, cylindric, straight or flexuous or twisted, thickening toward the top, white, silky, covered by white fibrils that bruise bluish, solid to tubular or fibrous, enlarged at base with many radiating white rhizomorphs that bruise bluish. Veil cobwebby, white, tearing to form a fragile annular zone dusted with purplish brown spores in the upper regions in maturity, soon disappearing if forming at all. Flesh white, bruising bluish. Odor and taste farinaceous, sometimes bitter. **MICROSCOPIC FEATURES:** Spore print violaceous dark. Spores 9–13 by 6–8 µm, nearly ellipsoid, thick-walled, wall 1 µm thick, yellowish brown, with a broad germ pore and short nipples at the ends. Clavate-ventricose, often constricted in the middle, hyaline. Pleurocystidia 17–33

Psilocybe hopii dramatically changes color as the maturing mushrooms produce purple brown spores.

by 4.5–11 µm, abundant, hyaline, ventricose-rostrate, sometimes lageniform. Cheilocystidia hyaline, 17–34 by 5–9 µm, similar to pleurocystidia 5.5–10 µm, lageniform or sublageniform, with a long, cylindrical, sinuous, single or branched neck, sometimes irregularly moniliform or twisted. Both types of cheilocystidia mostly sheathed with a hyaline gelatinous layer.

HABIT, HABITAT, AND DISTRIBUTION: First collected from the San Francisco Peaks region of north-central Arizona, Coconino County, directly north of Flagstaff. Now thought to be widely distributed across that plateau. Found in August and September. Scattered to gregarious on black soil with high content of organic material, in an aspen (*Populus tremuloides*) forest, with Douglas fir (*Pseudotsuga menziesii*) and limber pine (*Pinus flexilis*), in open places speckled with bracken ferns (*Pteridium aquilinum*) around 8000–10,000 feet (~2400–3000 m) in elevation. Given the broad ranges of these tree species, I suspect the distribution of *Psilocybe hopii* and the elevations at which it is found is much broader than presently reported.

COMMENTS: This species is named in honor of the Hopi people of the mesa region of north-central Arizona. *Psilocybe hopii* prefers pine forests, including ones recovering from the ravages of fire from decades before.[51] The phenotype bears a strong similarity to *Psilocybe cyanescens*, a low-altitude species. Experienced myconauts speculate that beavers propagate both of these species while building their dams and dens, since beavers drop trees and create wood chips nutritive to both species. When these dams break, myceliated wood spreads downstream. I think this hypothesis is highly credible. I wonder how many other Psilocybe species are associated with beavers and other wood-munching rodents.

Psilocybe ingeli B. VAN DER MERWE, A. ROCKEFELLER & K. JACOBS

CAP: 1–3 cm broad. Convex to hemispheric, broadly umbonate, with incurved to decurved margin, even, translucent striate to midway. Covered with a separable gelatinous pellicle when fresh. Caramel brown when moist and fresh, hygrophanous, fading to gray in drying, bruising bluish especially along the margin. **GILLS:** Sinuate, light gray at first, becoming dark brown with spore maturity, with whitish margins. **STEM:** 30–70 mm long by 2–6 mm thick, pruinose to covered with scattered fibrillose patches. Brown in color, adorned with whitish basal mycelium, bruising bluish. **MICROSCOPIC FEATURES:** Spores dark purplish brown, 7.5–9 by 5–7.5 µm, ovoid to ellipsoid. Cheilocystidia 11–21 by 4–8 µm, sublageniform to subclavate, tapering to snout-like at apices. Pleurocystidia scattered, 12–25 by 5–9 µm, clavate to sublageniform, narrowing toward the tips.

HABIT, HABITAT, AND DISTRIBUTION: Grows scattered on bovine dung in manured pastures in the late summer rainy season in South Africa.

COMMENTS: A rare species; published based on a lone specimen,[52] I include this species so more collections can be found and the species preserved. Its bluing reaction strongly suggests it is psilocybin-active. The morphology of this species is consistent with many psilocyboid species growing in pasturelands.

Psilocybe ingeli is a rare species growing in the pasturelands of South Africa.

Psilocybe ingeli is strongly hygrophanous and bruises bluish from handling or naturally from drying.

Psilocybe liniformans GUZMÁN & BAS = *Psilocybe liniformans* var. *americana* (GUZMÁN & BAS) STAMETS

CAP: 1.0–2.5 cm broad. Convex to broadly convex, sometimes broadly umbonate, but not acutely papillate. Surface smooth, viscid when moist from a separable gelatinous pellicle. Dull grayish ochraceous brown or slightly olivaceous, hygrophanous, fading in drying from the center, becoming straw brown, sometimes bluish green toned. **GILLS:** Attachment adnexed, close to distant, dark chocolate brown to purplish brown, with gelatinous and removable elastic edge. **STEM:** 14–30 mm long by 1–2 mm thick. Equal to swelled toward the base. Whitish to pale brownish, darker below, bruising bluish where injured, especially near the base and the apex. Surface pruinose above and finely fibrillose in the lower regions. Partial veil thin, soon disappearing. **MICROSCOPIC FEATURES:** Spores dark grayish purple brown in deposit, ellipsoid in both side and face views, 12–14.5 by 7.5–10 µm. Pleurocystidia absent. Cheilocystidia 22–33 by 5.5–9 µm, lageniform with an extended neck more than 6 µm long by 1.5–2.5 µm thick.

Psilocybe liniformans bruises bluish soon after handling the stem.

When using a pin, *Psilocybe liniformans* has separable gelatinous gill edges.

Psilocybe liniformans is found most frequently in horse pastures.

HABIT, HABITAT, AND DISTRIBUTION: Scattered to gregarious on horse dung or in soil enriched with manure in meadows and pastures. Growing in the late summer through autumn. The European form is known only from the Netherlands. *Psilocybe liniformans* var. *americana* has been collected in Washington, Oregon, and Michigan and also reported from Chile.[53] It is likely to be more broadly distributed in manure-enriched grasslands, but infrequently found compared to *Psilocybe semilanceata*.

COMMENTS: A distinct species, *Psilocybe liniformans* is becoming increasingly scarce for unknown reasons. As with its variety, *Psilocybe liniformans* var. *americana*, the whitish gill edge can sometimes but not always be peeled away especially when the mushrooms are fresh and wet. However, this varietal, *Psilocybe liniformans* var. *americana*,[54] may be more closely related and even conspecific to *Psilocybe semilanceata*, one of its closest phylogenetic relatives. The phenotype seen in my collections shows a much more robust fruitbody and thicker stem without a sharp papilla seen with typical *Psilocybe semilanceata*. Weakly to moderately active. Stijve and Kuyper found maxima in the European variety of 0.16 percent psilocybin, 0 percent psilocin, and 0.005 percent baeocystin.[55] Beug and Bigwood reported 0.59–0.89 percent psilocybin and 0 percent psilocin in the American form.[56] However, both analyses were based on single collections and details on drying and storage conditions are lacking. Hence, the species' true potency is likely underestimated. See also *Psilocybe callosa, Psilocybe hispanica, Psilocybe fimetaria*, and *Psilocybe semilanceata*.

Psilocybe mairei SINGER

CAP: 1.5–3.5 cm broad. Convex to campanulate to conic-campanulate, expanding with age, but not umbonate. Surface viscid when moist from a sometimes-separable gelatinous pellicle. Orangish brown, becoming olive toned, hygrophanous, fading in drying to yellowish white. Flesh amber, bruising bluish where injured. **GILLS:** Attachment adnate, pallid at first soon darkening, becoming purplish brown with the edges remaining whitish fringed. **STEM:** 25–75 mm long by 2–5 mm thick. Equal to slightly enlarged toward the base. Whitish to yellowish white, bruising bluish where injured. Surface pruinose above and finely fibrillose below. Stuffed with a silky pith. Base of stem adorned with thick, radiating white rhizomorphs. Partial veil cobwebby, white, sometimes leaving fibrillose remnants on the stem dusted purple brown with spores. **MICROSCOPIC FEATURES:** Spores dark purplish brown in deposit, elongate-ellipsoid, 10–12(13.5) by 5.5–7 µm. Pleurocystidia absent or near to gill edge. Cheilocystidia 30–40 by 6–8 µm. Variable in form: lageniform, fusiform, or ampullaceous.

Psilocybe mairei is potently psilocybin-active. Note bluish tones on cap in upper right specimen.

Psilocybe mairei could be the same psilocybin mushroom species depicted in the 7000-year-old Tassili Bee-Mushroom Man cave art found in southern Algeria (see page 25).

HABIT, HABITAT, AND DISTRIBUTION: Known only from Morocco and Algeria in North Africa from October to December. Growing gregariously on soil rich in woody debris, in forests mixed with pine (*Pinus pinaster*), fir (*Abies pinsapo*), and oak (*Quercus pyrenaica*). A few collections have been recently found at around 4200 feet (approximately 1300 meters) in elevation.

COMMENTS: Potent, judging by the bluing reaction and case reports, although no published analyses are known to me. This is the only wood-decomposing, bluing *Psilocybe* that I know from North Africa. Prior to the expansion of the Sahara, North Africa enjoyed a moister climate and undoubtedly hosted many more mushrooms than known today. *Psilocybe mairei* was probably much more common before the climate in this region dramatically changed. Given that it has likely persisted in this ecosystem for thousands of years, I wonder if the Indigenous people of that region long ago knew of this species. Today, they do. Judging by the rhizomorphic mycelia at the base of the stem, this species looks like a good candidate for creating satellite colonies.

Psilocybe makarorae P.R. JOHNSTON & P.K. BUCHANAN

CAP: 1.5–3.5 cm broad. Conic to campanulate, expanding to broadly convex, usually with a pronounced blunt umbo. Surface tacky to dry. Yellowish brown to orangish brown, fading in drying, lighter toward the margin, which is striate when moist. Flesh whitish, bruising greenish blue where injured. **GILLS:** Attachment adnexed, pale grayish brown with concolorous edges. **STEM:** 30–60 mm long by 2–4 mm thick, equal, surface adorned with patches of silky fibrils, readily bluing, white to brownish near the base, which radiates white rhizomorphs. Partial veil cortinate, leaving fibrillose veil remnants along the cap margin when young, soon disappearing, but not forming an annular ring on the stem. **MICROSCOPIC FEATURES:** Spores dark purplish brown in deposit, ellipsoid in side view, ovoid to subrhomboid in face view, 6.5–10 by 4.5–6.5 µm. Pleurocystidia similar to cheilocystidia, ventricose-rostrate to mucronate, 4–8 µm thick, with a simple neck, 2.5–4 µm long. Cheilocystidia 18–26 by 6–9 µm, ventricose-rostrate to mucronate with a short, simple neck, 3–5 µm long.

HABIT, HABITAT, AND DISTRIBUTION: Found scattered to gregarious on rotting wood and twigs of southern beech (*Nothofagus*), on the North and South Islands near lakes and picnic grounds in the vicinity of Makarora (Otago Lakes, Franz

Psilocybe makarorae is a potent psilocybin mushroom from New Zealand.

Psilocybe makarorae strongly bruises bluish where touched or injured.

Josef Glacier), New Zealand, in the fall. Probably more widely distributed.

COMMENTS: Potency unknown but probably moderately active given the bluing reaction. This species resembles *Psilocybe subaeruginosa* but is delineated by its darker brown color, smaller spores, and shorter pleurocystidia. The radiating rhizomorphs from the swollen stem base makes this species a good candidate for creating satellite colonies using stem butts as inoculants.

Psilocybe maluti grows on dung in the southern regions of Africa.

Psilocybe maluti B. VAN DER MERWE, A. ROCKEFELLER & K. JACOBS

Common Name: Koae-ea-lekhoaba

CAP: 0.5–2 cm broad by 2–4 cm in length. Secotioid in shape, umbonate to papillate with incurved to recurved margin, wavy but irregular, sometimes striated to midway, often adorned with appendiculate whitish partial veil remnants. Golden to caramel when adolescent, lightening in age but becoming darker grayish in the lower regions as spores mature. **GILLS:** Adnate, light brown at first, darkening to dark purple with maturity with edges creamy white. **STEM:** 40–80 mm long by 2–4 mm thick, cylindrical, surface fibrillose, whitish at first, becoming caramel brown with age, bruising bluish. **MICROSCOPIC FEATURES:** Spores dark purple to dark purple, brown to dark brown in mass, 13–15.5 by 8.5–10.5 μm, ovoid to ellipsoid. Cheilocystidia clavate, mucronate, 19–31 by 7–11 μm. Pleurocystidia absent.

Psilocybe maluti is psilocybin-active and is a secotioid species, meaning the gills are encased.

HABIT, HABITAT, AND DISTRIBUTION: Scattered, coprophilic, on the dung of herbivores of bovine and horses in the highlands of Lesotho and South Africa, November to February.

COMMENTS: A newly described species, this Psilocybe has a secotioid cap, meaning that the cap typically remains closed, limiting airborne

Psilocybe maluti is a secotioid fungus whose caps do not expand, limiting airborne release. Birds have been observed eating these mushrooms. One theory is that the mushrooms mimic fruit to attract birds who peck at them. Their beaks then will carry multiple spores, allowing for mating between spores when carried to distant new habitats.

dispersal, although gills producing spores are forming. Some theorize that these secotioid Psilocybes attract birds as their morphologies may imitate fruits. Crows and other birds have been observed pecking at these mushrooms by Indigenous people. Traditionally known and used by the Indigenous Basotho in the Kingdom of Lesotho, whose oral traditions suggest that this psilocybin mushroom had been used in spiritual ceremonies long before its recent identification by academic mycologists. The Basotho used these mushrooms for divinatory purposes. It is very likely that the use of this newly named species stretches far back in time, but being an oral tradition, it is not possible to document its duration of historical use. That academic mycologists have been unaware of this taxon suggests to me that its historical and original use must be ancient and independent of foreign influence. Two other secotioid species are *Psilocybe weraroa* and *Psilocybe gandalfiana* nom. prov. For more information, see pages 198 and 226.

Psilocybe medullosa (BRESADOLA) BOROVIČKA

CAP: 1–2 cm broad. Obtusely conic to campanulate, often with a low umbo, expanding with age to broadly convex. Orangish brown to yellowish brown to brown, darker toward the center, lighter along the margins, hygrophanous, fading to pale yellowish brown or grayish brown. Surface smooth, translucent-striate and viscid

Psilocybe medullosa grows in groups and is mildly psilocybin-active.

Psilocybe medullosa is a classic woodland Psilocybe: conic, with a translucent striate cap when moist and young.

Galerina species (left) can be deadly and have rusty brown gills and spores, whereas *Psilocybe* species, like *Psilocybe medullosa* (right), can have dark brown gills and brown spores.

Even to experts, *Psilocybe medullosa* looks like a *Galerina* species when viewed from above. When the gills and spores are viewed, these mushrooms be distinguished from one another. A mistake in identification can be deadly.

when moist from separable gelatinous pellicle. **GILLS:** Attachment adnate to adnexed, close to subdistant, moderately broad, with edges fringed with whitish edges. Color dull brown to rusty brown at maturity. **STEM:** 40–70 mm long by 2–3 mm thick. Equal to slightly enlarged at the base, brittle, tubular, and somewhat flexuous, with a base often tufted with whitish mycelium. Covered with whitish fibrillose patches, brownish beneath. Partial veil whitish, thin, cortinate, often soon obscure. **MICROSCOPIC FEATURES:** Spores dark purplish brown in deposit, 7–9 by 4–5 µm. Basidia clavate, 20–25 by 6–7 µm. Pleurocystidia absent. Cheilocystidia 20–40 by 4–7 µm, fusoid-ventricose to lageniform toward apex, attenuating into an elongated flexuous neck, 1–2 µm thick.

HABIT, HABITAT, AND DISTRIBUTION: Scattered to gregarious on detritus of conifer trees along trails and roads in Europe. Likely to be more widely distributed than presently recorded.

COMMENTS: This European species is geographically distant but genetically very close to the predominately North American species *Psilocybe silvatica*. It is very weak in its psilocybin content. They may be synonymized and share the same general aspect, being slender tall and with most fruitbodies having comparatively longer stems than the caps are wide. *Psilocybe medullosa* caps tend to expand to broadly convex to nearly plane, whereas the *Psilocybe silvatica* I have collected have caps that remain more conic to convex, but not plane. *Psilocybe medullosa* is a relatively rare species and may have a bluish tone at the base of the stem or in the basal mycelium. It is also similar to *Psilocybe atrobrunnea* and *Psilocybe pelliculosa*, which are also weakly psilocybin-active.[57]

Psilocybe mescaleroensis GUZMÁN, WALSTAD, E. GÁNDARA & RAM.-GUILL.

CAP: 2–6 cm broad. Convex to broadly convex, sometimes with a low, broad umbo. Yellowish brown, hygrophanous, with a translucent-striate margin when moist that is often undulating. Gelatinous pellicle separable when wet. Flesh bruising bluish where damaged. **GILLS:** Attached, adnate to adnexed. Cream color when young, darkening to chocolate brown with spore

Psilocybe mescaleroensis is a potent psilocybin mushroom that frequents the meadows interspersed within pine forests of New Mexico in the United States.

maturity. **STEM:** 5–10 cm long by 5–20 mm thick, white, sometimes adorned with faint tawny fibrillose patches, equal, fibrous, often enlarging toward the apex. Annulus whitish, bruising bluish, thinly membranous at first, fragile, often leaving a fibrillose annular zone with dark brown spores with maturity. Stem base adorned with radiating white rhizomorphs. Flesh and fibrils bruising bluish where damaged. **MICROSCOPIC FEATURES:** Spores dark chocolate brown, subrhomboid to subovoid, thick-walled, 9–13 by 6–8 µm. Pleurocystidia absent. Cheilocystidia 20–30 by 6 µm, fusiform or ventricose-rostrate, sometimes forked.

HABIT, HABITAT, AND DISTRIBUTION: Scattered, growing in the summer and fall on grassy grounds in meadowed ponderosa pine (*Pinus ponderosa*) forests, rich in buried wood debris. Also found in the subalpine spruce-fir forests of Sierra Mescalero in south-central New Mexico near Lincoln National Forest in Lincoln and Otero Counties, and collected in Colorado with one sighting in Kansas. This species often occurs near gopher holes, sometimes growing directly out of them. It is likely much more broadly distributed than presently reported.

COMMENTS: Thought to be a potent Psilocybe, *Psilocybe mescaleroensis* is named to honor the Mescalero Apache.[58] This species is most closely related to *Psilocybe hopii* and in the future these two taxa could be found to be the same species. This species also generally resembles *Psilocybe cyanescens* and *Psilocybe azurescens*, both of which also grow in close association with rodent, and sometimes beaver, activity. iNaturalist records numerous collections of this species since its first publication in 2007. Given the abundant rhizomorphs at the base of the stem, I think this species could be easily transplanted using the stem-butt inoculation technique described on page 78.

Psilocybe mexicana HEIM

Common Names: Nize (Mazatec), Pajaritos (Spanish; Little Birds)

CAP: 0.5–3.0 cm broad. Conic to campanulate to subumbonate, to convex with maturity, often with a small umbo. Surface viscid to smooth when moist, translucent-striate from the margin halfway to the disc. Margin sometimes decorated with fine fibrils. Brownish to deep orangish brown, hygrophanous, fading in drying to yellowish, becoming opaque, often with bluish tones from age or where injured. **GILLS:** Attachment adnate to adnexed, sometimes sinuate, pale gray to dark purplish brown with spore maturity, typically with whitish edges. **STEM:** 40–125 mm long by 1–3 mm thick, equal to narrowing toward the base, smooth, and hollow. Straw yellow to brownish, darkening with age or where injured. Partial veil thinly fibrillose, whitish, leaving fibrillose veil remnants on the upper regions of the stem. Flesh reddish brown, bruising bluish where injured. **MICROSCOPIC FEATURES:** Spores dark purplish brown to blackish purple brown

Psilocybe mexicana is a grassland species revered by Indigenous peoples in the highland meadows of Mexico and Guatemala. It is one of a few species that form sclerotia, which contains psilocybin but less than within the fruitbodies. Sclerotia typically form underground but can sprout into mushrooms. Sclerotia are easy to grow from the mycelium in darkness, but light inhibits their formation.

in deposit, ellipsoid to subellipsoid in side view, subrhomboid in face view, 8–12 by 5.5–8 µm. Pleurocystidia absent or similar to cheilocystidia when near the gill edge. Cheilocystidia 13–34 by 4.4–8.8 µm, fusoid-ampullaceous, sublageniform with abbreviated apices, 1.5–3.3 µm, occasionally forking.

HABIT, HABITAT, AND DISTRIBUTION: Solitary to gregarious in meadows, often in horse pastures, in soils rich in manure, along field–forest (deciduous) interfaces, most common at 3300–6500 feet (1000–1800 m) elevation. Found in June through September in subtropical Mexico (Michoacán, Morelos, Jalisco, Oaxaca, Puebla, western Xalapa, Veracruz) to Guatemala. The trees surrounding meadows hosting this species include sweetgum (*Liquidambar styraciflua*), oaks (*Quercus* spp.), alders (*Alnus* spp.), and others. Also found along river valleys in pasturelands sometimes bordered with sycamore (*Platanus lindeniana*).

COMMENTS: Moderately potent. Heim and Hofmann first detected 0.25 percent psilocybin and 0.15 percent psilocin. Fresh specimens are undoubtedly more potent. This mushroom forms sclerotia in culture that have less psilocybin than the fruitbodies. The sclerotia can be transplanted in soil and fruited (see pages 66–67).[59] I like to think of *Psilocybe mexicana* as the Liberty Cap of Mexico in that the two species share an affinity for habitats (manured grasslands), have similar shapes often with an umbo, are tall and thin, and have deeply translucent-striate margins when moist. Both species also have much more psilocybin than psilocin and typically do not bruise bluish, and if they do, rarely, it is more evident in the mycelium attached to the base of the stems. See also *Psilocybe tampanensis*, which may prove to be conspecific. More analyses are needed to disambiguate these closely related taxa.

Psilocybe muliercula SINGER & A.H. SMITH

Common Names: Little Women, Wasson's Psilocybe

CAP: 0.5–5 cm broad. Obtusely conic to conic-campanulate, then convex, expanding to broadly convex to nearly plane, often with an even to undulating, flaring margin. Surface smooth, shortly translucent-striate along the margin. Orangish brown to reddish brown to vinaceous brown, hygrophanous, lightening to a pale orangish light brown, often darkened with bluish streaks. Flesh white, readily bruising bluish where injured. **GILLS:** Attachment adnexed to ascending sinuate, close, brown, darkening to dark brown with spore maturity with edges whitish fringed. **STEM:** 20–100 mm long by 2–7 mm thick. Equal to enlarged toward the base, covered with fine fibrillose patches in the lower two-thirds and typically smooth above. Pale white to grayish brown, readily bruising bluish where injured or from handling. Flesh pinkish brown to pale white. **MICROSCOPIC FEATURES:** Spores dark violaceous brown, subellipsoid to ellipsoid-ovoid, 6–9.9 by 3.8–4 µm. Pleurocystidia absent. Cheilocystidia 16–24 by 3.5–6 µm, sublageniform, tapering with an elongated neck, 1.5–3 µm thick, sometimes forked and mucronate.

Psilocybe muliercula is a classic, potent, caramel-capped Psilocybe growing in Mexico. Its range is likely greater than currently reported.

HABIT, HABITAT, AND DISTRIBUTION: Gregarious to cespitose, in swampy soils, often in soil banks along streams, ravines, and roads in fir (*Abies*) and pine (*Pinus*) woodlands at elevations of 6890–11,480 feet (2100–3500 m), July through September. Dr. Gastón Guzmán collected this on the slopes of Nevado de Toluca, approximately 50 miles west of Mexico City, and Alan Rockefeller found this Psilocybe along a road near Puebla, Mexico. It is likely more widely distributed than reported in disturbed soils, especially along roads and paths through forested areas.

COMMENTS: This species was also described as *Psilocybe wassonii* by Roger Heim to honor R. Gordon Wasson, who worked with Heim on Mexican psilocybin mushrooms. However, Rolf Singer and Alexander H. Smith published the name *Psilocybe muliercula* a few weeks prior and with a Latin description, and thus their name took priority. These two factions of mycologists were highly academically competitive—and critical—of one another. In the mid-1950s, both groups contributed significantly to deepening scientific understanding of these species by studying dried specimens they had obtained from the Matlazinca Indigenous peoples near Tenango del Valle west of Mexico City.

Psilocybe natalensis grows in pasturelands and is a fairly potent psilocybin species. This photograph by Jochen Gartz is of the type collection.

Psilocybe natalensis GARTZ, REID, A.H. SMITH & EICKER

CAP: 1.4–6.0 cm broad. Obtusely conic, expanding with age to hemispheric to convex to broadly convex at maturity, occasionally with a small umbo. Golden to yellowish toward the disc, sometimes whitish overall, not hygrophanous, and often bluish tinged along the margin, which can be striate. Surface smooth to irregular in places. **GILLS:** Attachment bluntly adnate to subdecurrent in age. Buff at first, then dark purplish brown with whitish margins. **STEM:** 40–120 mm long by 2–10 mm thick. Surface smooth, silky white, straight to curved and enlarged near the base. Bruising bluish green where injured, particularly at the base. Partial veil white, bruising bluish, membranous, fugacious, sometimes absent, and leaving fibrillose veil remnants often decorated with purplish brown spores. **MICROSCOPIC FEATURES:** Spores deep violet in deposit, broadly ellipsoid in side view and ovoid in face view, 10–15 by 7–9.4 µm. Basidia 4-spored, although 1-, 2-, and 3-spored basidia observed. Pleurocystidia scattered to abundant, clavate to lanceolate, sometimes mucronate, 18–40 by 10–15 µm. Cheilocystidia lageniform, 16–22 µm long by 5.5–8 µm broad at the base, narrowing into a neck, 3.5–6 µm long, and then swelling at the apex.

HABIT, HABITAT, AND DISTRIBUTION: Scattered to gregarious on cow manure or in manure-enriched pastures. Likely to be found on the dung of other megafauna herbivores. Found in December to early March, and originally reported from Natal, South Africa,[60] at 4920 feet (1500 m) elevation; also reported east of Pretoria to the coastal regions. Likely to be more widely distributed across Africa.

COMMENTS: *Psilocybe natalensis* can resemble closely related Psilocybes and be difficult to separate macroscopically. They all are likely to have evolved from a common ancestor. This species is moderately to highly potent. Jochen Gartz estimated psilocybin and psilocin content to be similar to those of *Psilocybe cubensis*.[61] A white phenotype is sometimes collected, with a bluish membranous annulus. This species can also be easily cultivated.

Psilocybe niveotropicalis OSTUNI, ROCKEFELLER, JACOBS & BIRKEBAK

CAP: 0.9–6 cm broad. Convex, expanding to broadly convex to plane in maturity. Color pale white to light yellow to medium dark brown, at times with olive greenish hues, hygrophanous, gray to grayish white when fresh, golden yellow to orange to brown or white umbo, bruising bluish. Young specimens darker brown. Fruitbodies darkening in drying, sometimes bluish black. Surface smooth to subviscid, glabrous, translucent-striate near the margin when moist, margin decurved at first, even in age. Cap broadly umbonate but can be acutely umbonate when young, margin becomes split and irregular at maturity. Partial veil membranous, white covering the gills when young. **GILLS:** Close, attached, adnexed to subdistant with two tiers of intermediate gills. Whitish when young, soon light brown to dark purplish brown at spore maturity. **STEM:** 15–55 mm long by 1–9 mm thick, central, smooth, fibrillose-striate, cylindric, equal, somewhat subbulbous, base sometimes hypogeous, whitish to sorrel brown, solid or hollow, with white mycelium at the base. Adorned with a whitish membranous annulus, soon covered with purplish brown spores, rarely not present. **MICROSCOPIC FEATURES:** Spores dark purplish brown en masse, 7.8–9.8 by 6.8–8.6 µm, rhomboid or subrhomboid in face view, thick-walled. Basidia 22–30 by 6.0–10 µm, 2- or 4-spored, rarely 3-spored, hyaline, thin-walled, cylindric-vesiculose to subclaviform. Pleurocystidia 17–33 by 7.8–12.6 µm, hyaline, thin-walled, subventricose, fusoid-ventricose, scattered but more abundant near edge, occasionally with double apex, subgelatinous secretion at the apex. Cheilocystidia 9.3–26.2 by 5.8-11.4 µm, hyaline, thin-walled, subventricose, fusoid-ventricose, sometimes subpyriform, with an acute or obtuse apex, usually in clusters but sometimes scattered, sometimes with a subgelatinous secretion at the apex.

HABIT, HABITAT, AND DISTRIBUTION: Solitary, gregarious to cespitose-imbricate from January to June, then mid-September, on beds of wood-chip mulch in well-maintained landscaped neighborhoods typically under the shade of small shrubs. Known only from Palm Beach County, Florida. Likely to be more widely distributed.

COMMENTS: Moderately potent, estimated to be 1–2 percent of dried mass. This newly found, bluing, wood-loving species is remarkable for

Psilocybe niveotropicalis grows in southern Florida. Note the strongly hygrophanous cap.

Psilocybe niveotropicalis is estimated to be moderately potent in psilocybin content. Note the bluing reaction in specimen lower right.

its light color, membranous annulus, and preference for landscaping chips.[62] At first glance, *Psilocybe niveotropicalis* is morphologically similar to the dung-loving *Psilocybe cubensis*, but it has smaller spores and is ecologically distinct, growing from wood-chip mulch and not dung. That it occurs in decorative wood chips suggests that this species will become more common in residential landscaped habitats. Its natural origin has not yet been determined. Like many of the *Psilocybe* species that appear in wood mulch, the mycelium may reside within host trees and only appear when detritus—or wood chips—are created, much like *Psilocybe cyanescens*.

Psilocybe ochraceocentrata C. SHARP & A. BRADSHAW

CAP: 65–75 mm broad, convex, expanding to plane in age, often with an undulating margin, which is downcurved when young, often cracking along the edge with maturity. Cream colored when young, then vinaceous gray becoming more ochraceous to fulvous toward the center, fading to pale yellow near the margin. Surface radially silky and having a dull sheen. Flesh bright cream colored, strongly bruising bluish green. **GILLS:** Adnate, few to crowed, pale at first, soon grayish-sepia to brown-vinaceous to very dark sepia from spores and almost black near margin. Gill surface differentially mottled, up to 12 mm deep, thin, papery, and fragile with smooth or finely scalloped edges. **STEM:** 20–95 mm long by 9–10 mm thick, cylindrical, hollow, sometimes longitudinally twisted, swollen at the apex with white tomentose tufts of mycelium attached to the base. Cream colored, readily bruising bluish. Partial veil thinly membranous, fragile, striate, forming an evanescent membranous ring, surface often adorned with purplish brown spores, soon disintegrating into a dark fibrillose annular zone, often disappearing with age. **MICROSCOPIC FEATURES:** Spores ellipsoid to lens shaped, dark vinaceous gray to purplish gray, variable in size, 9–13 by 8.5 µm, thick-walled, smooth with an apical germ pore. Cheilocystidia present. Pleurocystidia not observed in type collection.

HABIT, HABITAT, AND DISTRIBUTION: Often growing in clusters, from granitic sand in miombo and mixed deciduous woodlands, frequently with leaf litter attached. Also reported from the dung of several megafauna herbivores, including horses, rhinoceroses, and zebras, groups can stand up to 105 mm tall. Found in South Africa (Kosi Bay, Kwazulu-Natal to Port Alfred in the Eastern Cape), Harare Province, Zimbabwe, and likely to occur in Angola, Botswana, Burundi, Democratic Republic of Congo, Malawi, Mozambique, Namibia, Tanzania, and Zambia. I suspect this species is underreported and likely has a much broader distribution.

Psilocybe ochraceocentrata looks *very* similar and is closely related to *Psilocybe cubensis*.

COMMENTS: Often robust in form, this newly named species resembles *Psilocybe cubensis* and grows on cow, horse, zebra, buffalo, and rhinoceros dung, but also has been found in manure-enriched soils, sandy soils, in woodlands, and on decomposing forest detritus. *Psilocybe ochraceocentrata* could be an ancestor of *Psilocybe cubensis*. Recent analysis on the evolution of Psilocybes in this clade suggests that the progenitors of *Psilocybe ochraceocentrata* and *Psilocybe cubensis* may have originated in woodland ecosystems, and more recently, that is, in the past few million years and in the case of *Psilocybe cubensis,* jumped to dung.[63] One analysis showed combined psilocybin and psilocin to be about 0.5 percent, but it's likely to be much higher.

Psilocybe ovoideocystidiata GUZMÁN & GAINES

CAP: 1–8 cm broad. Convex to broadly convex, expanding in age to nearly plane, often with a low, broad umbo, with an even, sometimes slightly wavy margin. Dark chestnut brown when young, becoming more orangish brown to yellowish brown in color, hygrophanous, fading to beige cream. Flesh bruising bluish to bluish green. Older specimens can be nearly black. Translucent-striate along the margin when moist. **GILLS:** Attachment adnate to adnexed, with two tiers of intermediate gills. Pallid when young, becoming brownish as spores mature, edges whitish fringed. **STEM:** 15–130 mm long by

Psilocybe ovoideocystidiata is a potent psilocybin mushroom that typically grows naturally with box elders in the eastern United States. It is one of the few lignicolous species that fruits in the spring. Most wood chip–loving psilocybin mushrooms in low-elevation temperate regions appear in the fall.

2–20 mm thick, equal, hollow, base glabrous near the top and typically adorned with a fragile, fugacious membranous ring that soon deteriorates into persistent fibrillose ring in the mid regions, below which fibrillose patches form, colored whitish to irregularly yellowish, brownish, with or without bluish tones. **MICROSCOPIC FEATURES:** Spores dark purplish brown, 7–12 by 5.5–8.5 µm, rhomboid to subrhomboid in face view, subellipsoid in side view, thick-walled, 0.8 to 1.5 µm thick. Pleurocystidia globose-pyriform, narrowing at the base and apex, 20–40 by 12–16 µm. Cheilocystidia present, lageniform.

HABIT, HABITAT, AND DISTRIBUTION: Fruiting in the spring, from April to June. Often gregarious, increasingly widely distributed as hunter-gatherers spread this species, which readily adapts to newfound habitats. First found and common in Montgomery County in western Pennsylvania within the Ohio River Valley, preferring moist rural bottomlands, especially near dead or dying boxelder (*Acer negundo*), a species of maple, but also associated with the invasive Japanese knotweed (*Reynoutria japonica*). Thriving in wood-enriched disturbed soils,

Psilocybe ovoideocystidiata can form large colonies and be easily naturalized to create substantial fruitings like this one.

especially where there are washouts from streams overflowing and then receding. Growing in constructed wood mulch beds and where wood chips are used for landscaping. *Psilocybe ovoideocystidiata* is now reported north to New England, south to Mississippi, and from Washington State to Southern California. There are also reports from Germany and Switzerland. This species is spreading rapidly, likely augmented by avid cultivators who are establishing satellite patches.

COMMENTS: First discovered in the Ohio River Valley by Richard Gaines, who originally proposed that this species be called *Psilocybe vernalis* nom. prov. This lignicolous species is unique in that it prefers to fruit in the spring; most wood chip-loving *Psilocybe* species fruit in the fall. This species is easy to cultivate by transplanting stem butts or lenses of mycelium to areas rich in deciduous wood debris. Fruitings can be large, with hundreds of specimens forming. The name *ovoideocystidiata* comes from its fat pleurocystidia.

Psilocybe pelliculosa (A.H. SMITH) SINGER & A.H. SMITH

CAP: 0.5–2(3) cm broad. Conic becoming conic-campanulate with age, and rarely further expanding without splitting the cap margin, which is translucent-striate when moist. Dark chestnut brown to brown when moist, hygrophanous, fading to dark dingy yellow to pale yellow with tinges of light bluish green tones in drying. When wet, the surface is smooth, viscid from a thick, easily separable gelatinous pellicle. Flesh very thin, similar color as the cap. **GILLS:** Attached, ascending but adnate to adnexed, finally nearly seceding in age, close, narrow to moderately broad. Color dull cinnamon brown. **STEM:** 30–80 mm long by 1–2.5 mm thick. Equal above, swelling at base. Surface adorned with fine grayish fibrils, often powdered near apex. Dingy to gray, darkening to more brownish toward the base, inconsistently bruising a faint blue green. Darkening as the mushrooms dry. Partial veil

Psilocybe pelliculosa, the Trail Psilocybe, is low in psilocybin but can be found in prolific groups, especially where trails and roads are built in coniferous woodlands in the Pacific Northwest of North America.

Psilocybe pelliculosa tends to fruit late in the fall season, preferring colder, near-freezing days. I find this species in late October through December in Oregon, Washington, Idaho, and British Columbia. It sometimes bruises bluish at the base of the stem, but bruising is more commonly seen in the basal mycelium.

Psilocybe pelliculosa is so named for its easily peelable, separable, gelatinous pellicle. Note that the pellicle is not peelable if the mushrooms dry out. This feature is not exclusive to all psilocybin mushroom species, nor to the genus *Psilocybe*.

obscure or absent. **MICROSCOPIC FEATURES:** Spores 9–13 by 5–7 µm. Pleurocystidia absent. Cheilocystidia 17–36 by 4–7.5 µm, fusiform, attenuating to a long neck, 1.5–2 µm.

HABIT, HABITAT, AND DISTRIBUTION: In groups, sometimes scattered, on well-decayed conifer mulch. Most frequently found along paths in Douglas fir (*Pseudotsuga menziesii*) forests and along abandoned roads. Primarily a cold-weather species, I find it in mid to late fall to early winter. Reported throughout the Pacific Northwest as far south as San Francisco, California, to northern British Columbia, Canada, and east to Coeur d'Alene, Idaho.

COMMENTS: Very low in psilocybin content. I am always delighted to find this species. I have seen thousands of them along infrequently used roads cut through forests. This is a trail-following species. Once I found *Psilocybe pelliculosa* deep in an old-growth forest directly at the base of a several-hundred-year-old Douglas fir tree distant from any human impact—the only sighting of its kind that I know of. Its mycelium is likely widespread in Douglas fir forests, and it fruits in disturbed soils rich with wood detritus. The conic cap, the gregarious nature of its colonies, the scattered fibrillose patches on the stem, and the inconsistent bluing reaction at the base of the stems are some of the most distinctive features of this species. *Psilocybe pelliculosa* is nearly identical to *Psilocybe silvatica* and to a lesser degree *Psilocybe medullosa*. This mushroom is generally very similar to conic-capped *Hypholoma* species, especially to *Hypholoma dispersum* (= *Naematoloma dispersum*) and *Hypholoma ericaeum*, but these lack the separable gelatinous pellicle characteristic of *Psilocybe pelliculosa* and never bruise bluish.

Psilocybe samuiensis GUZMÁN, BANDALA & J.W. ALLEN

CAP: 0.7–1.5 cm broad. Convex to conic-convex to campanulate, often umbonate with a small papilla. Surface translucent-striate near the gill edge and viscid when moist from a separable gelatinous pellicle. Chestnut to reddish brown to straw color when young, strongly hygrophanous, fading in drying to straw color or brownish clay. **GILLS:** Attachment adnate, clay colored, then violaceous brown to chocolate violet with whitish margins. **STEM:** 40–65 mm long by 1–2 mm thick. Whitish to yellowish and covered with fibrillose sheath of veil remnants. Equal to slightly enlarged toward base. Concolorous with cap, bruising bluish where injured. Partial veil cortinate, leaving a superior, fibrillose annular zone that soon disappears. **MICROSCOPIC FEATURES:** Spores purplish brown in deposit, rhomboid to subrhomboid, 10–13 by 6.5–8.0 µm. Pleurocystidia scattered, ventricose toward base, sublageniform toward the apex, 16–20 by 5–6.4 µm. Cheilocystidia ventricose at base, lageniform, narrowing to a thinner neck, often forked, 18.5–28(30) by 5–7(8) µm.

HABIT, HABITAT, AND DISTRIBUTION: Growing in well-manured, clay-like soils in pastures, meadows, or among rice paddies. First reported by John W. Allen from Koh Samui Island off Thailand, *Psilocybe samuiensis* is now known from other regions of Thailand, India, Malaysia, Borneo, Java, Indonesia, Bali, Vietnam, and possibly China. This mushroom has been collected during the summer. *Psilocybe samuiensis* may be widely distributed throughout the Sino maritime region.

COMMENTS: *Psilocybe samuiensis* is a slender mushroom with a general morphological similarity to the many grassland species that resemble *Psilocybe semilanceata* and *Psilocybe mexicana*.

Psilocybe samuiensis was first reported from Thailand and is now known to be widespread across Asia.

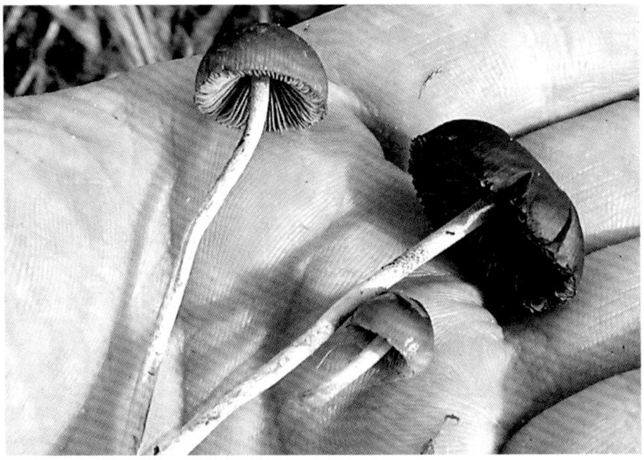

Psilocybe samuiensis is a classic grassland Psilocybe. If bruised, the flesh shows bluish tones.

Psilocybe samuiensis prefers a common habitat with *Psilocybe cubensis* and the groups of coprophilic *Copelandia* species centering around *Panaeolus cyanescens*. Active with psilocybin (up to 0.73 percent), psilocin (up to 0.52 percent), and baeocystin (up to 0.05 percent).[64] Stijve et al. reported up to 0.075 percent psilocybin and 0.60 percent psilocin.[65,66]

Psilocybe semilanceata (FRIES) KUMMER

Common Name: Liberty Cap

CAP: 0.5–2.5 cm broad. Conic to obtusely conic to conic-campanulate to campanulate, typically with a bluntly obtuse or acute umbo. Some specimens lack an umbo. The caps remain conic, rarely expanding to convex, but not plane. Margin translucent-striate when moist, incurved and sometimes folded-undulated in young fruitbodies, often darkened by spores along the edges. Color variable, extremely hygrophanous. Usually dark chestnut brown when moist, soon drying to a light tan or yellowish and occasionally with an olive tint. Surface viscid when moist from a separable translucent gelatinous pellicle. **GILLS:** Attachment mostly adnexed, ascending, close to crowded, narrow. Color pallid at first, rapidly becoming brownish and finally purplish brown with the edges remaining pallid. **STEM:** 30–100 mm long by 0.75–2(3) mm thick. (Stem length is highly variable and often longer in taller grass.) Slender, equal, flexuous, pliant. Pallid to more brownish toward the base. Surface smooth overall. Context stuffed with a fibrous pith. Partial veil thinly cobwebby, rapidly deteriorating, sometimes leaving a fragile annular zone of fibrils, usually darkened by purplish brown to black spores. **MICROSCOPIC FEATURES:** Spores dark purplish brown in deposit, ellipsoid, 12–14 by 7–8 µm. Pleurocystidia few to absent. Cheilocystidia 18–35 by 4.5–8 µm, lageniform with an extended and flexuous neck, often forked.

HABIT, HABITAT, AND DISTRIBUTION: Scattered to gregarious in pastures, fields, lawns, and other grassy areas in the fall, especially rich grasslands grazed by sheep and cows, though sometimes in horse pastures. The mushroom mycelium tightly

Psilocybe semilanceata, the Liberty Cap, is widespread in temperate grasslands of the world. Occasionally, the mycelium attached to the bases of the stems can show a bluing reaction. High in psilocybin but low in psilocin, this is a potent species that stores well compared to many other species.

Psilocybe semilanceata can be short or long stemmed, greatly influenced by the height of the grass with which it grows. The caps can variably have sharp or blunt nipples.

engulfs the bases of grasses and is suggested to be an endosymbiont.[67] Especially abundant in or about clumps of grass and sedge in the damper parts of fields, especially near ponds or swampy areas that partially submerge in heavy rains. Found in the Pacific Northwest of North America, Newfoundland of eastern Canada, northern Europe (Italy to Switzerland to Holland, Norway to France), in grasslands of South Africa, Chile, and northern India. Although reports are limited and difficult to document, this species likely grows in many temperate grassland regions of the world, including temperate pastoral regions of China and Russia. Its strong association with domesticated cattle and sheep suggests that this species has spread through these animals and the fodder grasses that were imported with them. Whether or not *Psilocybe semilanceata* is native to North America, prior to the arrival of Europeans, is an unanswered question, as several sheep species are indigenous to North America.

COMMENTS: A potent pastoral psilocybin species that is relatively easy to identify—despite that it does not typically bruise bluish upon injury. The unique nipple on the conic cap and the Liberty Cap's very long stem are two prominent features. This species was given its common name in the late 1700s because it resembles the emblem signifying freedom in the French Revolution. The Druids of Great Britain use this species sacramentally today, and many believe these mushrooms have long been a part of their spiritual practices (see pages 29–30). *Psilocybe semilanceata* grows prolifically around Stonehenge in Salisbury, England, which was built by the Druids as a place of worship.

Although petite in size, *Psilocybe semilanceata* is one of the more potent psilocybin-active mushrooms. Typically, 10 to 40 specimens are consumed, a range that is highly dependent on factors affecting potency. The high psilocybin and low psilocin content contribute to this species having a long shelf life. Bradshaw et al. reported a psilocybin content ranging from 3.33 to 15.77 mg/g (= 0.33 to 1.57 percent by dry weight) and a psilocin content ranging from 0.1 to 3.88 mg/g (0.1 to 0.38 percent by dry weight).[68] Jochen Gartz reported an average of 1 percent psilocybin, with a range from 0.2 to 2.37 percent. In separate studies, Gartz reported up to 0.98 percent psilocybin[69] and Stijve and Kuyper found a high psilocybin concentration (in one specimen) of 1.70 percent and a relatively high baeocystin content (0.36 percent).[70] Christiansen et al. found that psilocybin in dried specimens varied from 0.17 to 1.96 percent, with the youngest specimens being most potent on a dry weight basis.[71] Gartz also reported that psilocybin content was not adversely affected by the drying process. Ohenoja et al. found up to 0.87 percent psilocybin in recently dried specimens.[72] Herbarium samples collected more than 11 years prior contained 0.84 percent psilocybin, whereas those collected more than 33 years prior had 0.67 percent.[73] *Psilocybe semilanceata* holds its psilocybin content better than most psilocybin-active species. Psilocybin is a stable molecule compared to psilocin. The petite size of

Psilocybe semilanceata typically grows scattered in cow and sheep pastures and, much less frequently, where only horses graze. I have also found it in grassy areas without manure, but manure-enriched pastures clearly support more prolific fruitings.

this mushroom and its thin flesh aid in it quickly drying, which may also contribute to the longevity of its psilocybin content.

Since *Psilocybe semilanceata* is high in psilocybin but low in psilocin, it rarely bruises bluish. When it does bruise, the bluing reaction is most evident in the basal mycelium and, at times, the drying caps can have streaks of bluish green tones, but not typically so.

Several distinct varieties of *Psilocybe semilanceata* can be encountered over a lifetime of collecting. The archetypal *Psilocybe semilanceata* is easy to recognize macroscopically. However, its many forms can be confusing. Most forms have conic to campanulate caps with a blunt to sharp umbo. Some specimens lack an umbo. I have never found any that expand to broadly convex without the cap margins splitting from stress.

Notably, wood-decomposing and terrestrial *Galerina* species can coexist in the same habitat as grassland *Psilocybe* species, especially when a wooded area has been recently converted to lawns, pastures, or grasslands. I have collected woodland *Psilocybe* species on soccer and football fields many years after they were converted to lawns, although the fruitings declined over time and eventually became scarce to absent without contributions from the dung of cattle and sheep or other fertilizers.

Psilocybe callosa and *Psilocybe liniformans* are similar species but lack the distinctive umbo typical of *Psilocybe semilanceata*. *Psilocybe liniformans* var. *americana* may be more closely related to *Psilocybe semilanceata* than to *Psilocybe liniformans*. *Psilocybe semilanceata* is to temperate grasslands what *Psilocybe mexicana* is to subtropical grasslands. An interesting ecological study by Keay and Brown illustrated the close relationship of *Psilocybe semilanceata* to the rhizomes of grasses.[74]

Several secotioid collections once thought to be abberant forms of *Psilocybe semilanceata* may be new species, yet to be named. The coprophilic *Protostropharia semigloboides* group includes several indicator species that often grow close by.

Psilocybe serbica is prolific throughout much of Europe.

Psilocybe serbica MOSER & HORAK

CAP: 1–4 cm broad. Bluntly conic expanding in age to campanulate or convex, to broadly convex and plane at full maturity, typically retaining a broad umbo. Surface viscid when moist, but pellicle typically not separable. Margin translucent-striate when moist, opaque upon drying, even, can uplift in age; if so, irregular. Strongly hygrophanous. Surface brown to caramel to dingy orangish brown and pale straw yellow, becoming light yellowish to whitish upon drying, often with bluish stains. Flesh whitish to cream, typically bruising blue when injured. **GILLS:** Attachment ascending, adnate to adnexed to subdecurrent; light brown when young, darkening to purplish brown with spore maturity. **STEM:** 20–80 mm long by 1.5–8 mm thick, equal, enlarging toward the base. Surface covered with silky grayish fibrillose patches, sometimes bluish tinged. Partial veil white, finely cortinate, if evident at all, disappearing with maturity. Whitish rhizomorphs attached to stem base. **MICROSCOPIC FEATURES:** Spores dark purplish brown in deposit, ellipsoid, 10–13 by 5.5–7 µm. Pleurocystidia absent or few. Cheilocystidia 20–36 by 6–11 µm, lageniform with elongated, tapering necks, 1.5–2.5 µm thick.

Psilocybe serbica is a common European species and potently psilocybin-active. Note faint bluing on stems. Mushroom in upper right is NOT *Psilocybe serbica*.

HABIT, HABITAT, AND DISTRIBUTION: Often gregarious, in mixed woods, growing in groups in soils enriched with decomposing twigs, leaves, and wood debris from deciduous and conifer trees, often with stinging nettles (*Urtica*) and blackberries (*Rubus*). Usually found along paths, dirt roads, and ravines and along creeks, often in association with European beech (*Fagus sylvatica*). Sometimes found fruiting from twigs. Fruiting from mid-September into December, from Morocco to Spain and throughout central Europe to Greece, with one report from Russia.

COMMENTS: This is a classically shaped and potent *Psilocybe* with freshly air-dried specimens having nearly 2 percent combined psilocybin and psilocin.[75] Its strongly hygrophanous cap readily exhibits a bluing reaction upon injury or drying. This species appears to have a long-time association with native woodlands and could have been

A *Psilocybe serbica* colony shows fresh, dark forms and drying, lighter (hygrophanous) forms.

used by Indigenous peoples in the region due to its distinctive appearance, prevalence, easy-to-recognize bluing reaction, and its penchant for growing along trails that humans travel. Once infrequently reported, collections of this species are increasingly being posted on iNaturalist by experienced myconauts. As with many *Psilocybe* species, once *Psilocybe serbica is* discovered and harvested, it will be spread to new habitats by humans. I expect this species to become increasingly more common. Although rarely reported from wood chips—the preference of *Psilocybe cyanescens,* for instance—I think this species is likely to make the jump from forest soils to landscape mulching enriched with leaf litter due to human engagement and attempts to naturalize the species via transplanting or stem-butt transfers. Several other species are now known to be synonymous with *Psilocybe serbica*, including *Psilocybe bohemica, Psilocybe arcana, Psilocybe moravica*, and possibly *Psilocybe mairei*, which has also been reported from the Morocco region.

Psilocybe silvatica (PECK) SINGER & A.H. SMITH

CAP: 0.8–2.5 cm broad. Obtusely conic to campanulate and often with an acute umbo. Tawny dark brown when moist, fading to pale yellowish brown or grayish brown. Surface smooth, viscid when moist from a thin gelatinous pellicle that is barely separable, if at all. **GILLS:** Attachment adnate to adnexed, close to subdistant, narrow to moderately broad. Color dull rusty brown at maturity. **STEM:** 20–80 mm long by 1–3 mm thick. Equal to slightly enlarged at the base, brittle, tubular, and somewhat flexuous. Pallid to brownish beneath a whitish fibrillose covering. Partial veil poorly developed, cortinate, thin to obscure, and soon absent. **MICROSCOPIC FEATURES:** Spores dark purplish brown in deposit, 6–9.5 by 4–5.5 µm. Basidia 4-spored with some 2-spored as well. Pleurocystidia absent. Cheilocystidia 20–40 by 4.4–7(8.8) µm. Fusoid-ventricose to lageniform toward apex, attenuating into a long flexuous neck, 1.6–2.2 µm thick.

Psilocybe silvatica is weakly psilocybin-active and very similar to *Psilocybe pelliculosa,* except it has smaller spores and the cap tends to be a lighter brown.

HABIT, HABITAT, AND DISTRIBUTION: Gregarious but not cespitose on wood debris, wood chips, or in well-decayed conifer-enriched soils in the fall, mostly along infrequently travelled roadsides or trails. Known from the eastern United States from Michigan to New York to Ontario, Canada, and the Pacific Northwest of North America. Also reported from northern Europe.

COMMENTS: Contains small amounts of psilocybin and/or psilocin. Rarely found in sufficient quantities for journeying. Differing from *Psilocybe pelliculosa* in the length of the spores and cheilocystidia. The colonies of *Psilocybe silvatica* that I have found exhibit more yellowish tones in the caps compared to the nearly identical forms of *Psilocybe pelliculosa*. The European *Psilocybe medullosa* and *Psilocybe atrobrunnea* (= *Psilocybe turficola*) are also similar in appearance. See also *Psilocybe cyanofibrillosa*.

Psilocybe stametsii DENTINGER & FURCI

CAP: 1–1.5 cm broad. Conic to campanulate, distinct and pronounced acute umbo, margin undulates with maturity. Brown to dark yellowish ochre. Surface smooth, viscid when wet with a deep-reaching translucent-striate margin. **GILLS:** Subdistant, ascending, adnexed to sinuate, brownish cream initially, light brown in aging, with two sets of intermediate gills. **STEM:** 30–40 mm long by 1–2 mm thick, equal to flexuous, hollow in age, pruinose, dark brown, darkening to nearly black at base, lightening to red to pale orange toward the apex, bruising bluish, especially evident at stem base. **MICROSCOPIC FEATURES:** Spores smooth, small, subrhomboid to rhomboid with slightly thickened cell walls, with an apical germ pore, translucent brown under the microscope, 5.2–5.5 by 3.8–4.4 µm. Cystidia present, lageniform to utriform (leather bottle shaped), cell walls not thickened, hyaline in 5 percent KOH.

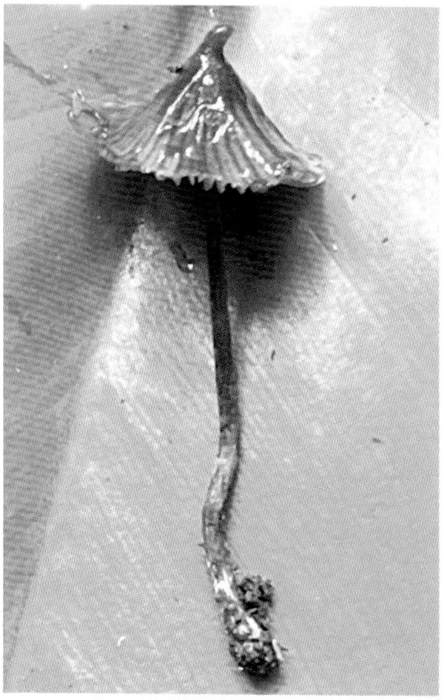

Psilocybe stametsii is a rare, strikingly nippled, bluing Psilocybe found in the high-altitude cloud forests of Ecuador.

HABIT, HABITAT, AND DISTRIBUTION: Solitary, growing in soils among decaying leaves along a trail in the Choco Andino cloud forest of Ecuador in Los Cedros Biological Reserve at an elevation of approximately 4600 feet (1400 m). Found in October and January.

COMMENTS: Uniquely shaped, with a pronounced wagging nipple on the cap. This is a rarely encountered species—thus far, only two solitary specimens have been found. Bryn Dentinger and Giuliana Furci honored me by naming this species *Psilocybe stametsii*. DNA analysis places it closest to *Psilocybe yungensis*. The shape of its pronounced nippled cap also resembles that of *Psilocybe hoogshagenii*. I hope more collections are found and cultures made to preserve this species and better understand its interrelationship to other species in the Ecuadorian cloud forests.

Psilocybe stuntzii GUZMÁN AND OTT

Common Names: Stuntz's Psilocybe, Stuntz's Blue Legs, Blue Ringers

CAP: 1.5–4(5) cm broad. Cap obtusely conic at first, soon expanding to convex to broadly convex-umbonate to nearly flattened or plane with the margin uplifting in very mature fruiting bodies. Margin translucent striate halfway to the disc when moist; decurved, then straightening, and finally elevated, undulating, and often eroded in extreme age. Dark chestnut brown, lighter toward the margin, which is often olive greenish; hygrophanous, fading to a more yellowish brown to pale yellow in drying. Some varieties tend to be more olive yellowish brown and are not very hygrophanous. Context relatively thin, watery brown or nearly concolorous with the cap. Surface viscid when moist from a

Psilocybe stuntzii is an easy to identify Psilocybe that is common in landscaping bark, typically from conifer trees. When young, it features a whitish, membranous partial veil.

The partial veil of *Psilocybe stuntzii* tears from the cap margin as each mushroom matures, falling to leave a fragile membranous annulus, usually bluing, in the upper two-thirds of the stem.

separable gelatinous pellicle. **GILLS:** Attachment adnate to adnexed, close to subdistant, moderately broad, with three tiers of intermediate gills. Color pallid in young fruiting bodies, soon becoming more brownish and eventually very dark brown with spore maturity. **STEM:** 3–6 cm long by 2–4 mm thick. Subequal, slightly enlarged at the apex and often curved and inflated at the base. (In some specimens, the stem will be extremely contorted in a "pig's tail" fashion.) Dingy yellow to pale yellowish brown. Surface dry, covered with pallid appressed fibrils to the annulus, and smooth above. Context stuffed with a fibrous whitish pith. Partial veil thinly membranous, typically streaked bluish, leaving a fragile membranous annulus as the cap expands, which soon deteriorates into a fairly persistent annular zone darkened by spores. Stem often with rhizomorphs protruding about the base. **MICROSCOPIC FEATURES:** Spores dark purplish brown to dark violet brown in deposit, rhomboid to subrhomboid in face view, ellipsoid in side view, 8–13.5 by 5.5–7.5 µm. Cheilocystidia lageniform, fusoid-ampullaceous, fusiform-lanceolate, 22–30 by 4.4–6.6 µm with an elongated and flexuous neck, 1.2–2 µm. Pleurocystidia absent.

HABIT, HABITAT AND DISTRIBUTION: Growing gregariously or in clusters on wood chips or in decayed conifer substratum, also in lawns and fields, in the fall to early winter and in the spring. Reported from western Oregon, Washington, and British Columbia. Abundant throughout the Puget Sound area of Washington State. This mushroom is very common in urban and suburban areas around institutions, commercial buildings, and dwellings.

COMMENTS: *Psilocybe stuntzii is* one of the weaker of the bluing Psilocybes. This species often grows in colonies of great numbers, in conifer bark used

This cluster of *Psilocybe stuntzii* does not show bluing on the annuli, which is usually dusted with dark purplish brown spores as the mushrooms mature.

for landscaping. *Galerina marginata* (= *Galerina autumnalis*), a deadly poisonous mushroom, can co-occur with *Psilocybe stuntzii* and has the same overall appearance to non-experts, except that it tends to be more orangish brown and has rusty brown spores (see page 13). Another variety grows in fields and is thinner in form. The *Psilocybe stuntzii* group encompasses a great variety of forms growing in varied habitats. Named in honor of Dr. Daniel Stuntz, who made the type collections along with Jonathan Ott at the University of Washington, Seattle. I contributed isotypes.

Psilocybe subaeruginascens HÖHNEL = *Psilocybe aerugineomaculans* (HÖHNEL) SINGER & A.H. SMITH

CAP: 1.0–5.8 cm broad. Conic to convex or campanulate to broadly subumbonate, but not papillate, eventually broadly convex to plane and uplifting in age. Surface viscid when moist, translucent-striate along the margin, soon drying. Orangish brown to olive brown to gray greenish brown, hygrophanous, fading in drying to dull yellow orange to straw colored. Flesh white to concolorous with cap, soon bruising bluish where injured. **GILLS:** Attachment broadly adnate to adnexed, sometimes decurrent, crowded, sometimes forking. Grayish brown to yellowish brown and eventually dark brown with spore maturity, often slightly mottled. Edges concolorous, bruising bluish where bruised. **STEM:** 30–60 mm long by 1.5–3.0 mm thick. White, bruising bluish, equal to slightly enlarged near base, which is often adorned with radiating white rhizomorphs that bruise bluish when injured. Surface and flesh whitish to concolorous with the cap, soon bruising bluish. Partial veil membranous, well developed, leaving a persistent membranous annulus in the median to superior regions of the stem, white until bruised, then bluish, usually dusted with purplish brown spores. Torn and fragile in age. **MICROSCOPIC FEATURES:** Spores dark purplish brown to dark violet brown in deposit, rhomboid to subrhomboid in face view, subellipsoid in side view, 7.7–12 by 6.6–8.5 µm. Basidia 4-spored, rarely 1-, 2-, or 3-spored. Pleurocystidia fusoid-ventricose with a blunted end, 2.2–3.3 µm thick. Cheilocystidia fusoid ventricose to sublageniform, 16–33 by 4.4–5.9 µm, with a neck 2.5–4 µm thick.

Psilocybe subaeruginascens is a potent psilocybin-active species widespread throughout Asia.

HABIT, HABITAT, AND DISTRIBUTION: Growing gregariously to cespitose on soils enriched with woody debris, in wood chips, and in wood chips mixed with horse dung; frequently found along trails or roadsides bordering deciduous forests. Fruiting April to July in temperate regions of South Korea, Japan, Malaysia, Indonesia, and South Africa. This species is probably more widely distributed than presently reported.

COMMENTS: A remarkable species, this membranous ringed *Psilocybe* looks uniquely different from most other lignicolous taxa in the genus. The bluish stains on the white stems are often pronounced, a reaction from handling. The whitish rhizomorphs adorning the stem base are of the transplantable type, meaning the stem butts of the mushrooms could be used to create new colonies into other wood-chip beds. This squat, collyboid, and membranously annulate Psilocybe is uniquely Asian and not yet known elsewhere. This species is likely in a center of a constellation of new species yet to be studied. See also *Psilocybe niveotropicalis* and *Psilocybe ovoideocystidiata* which are generally similar in appearance macroscopically but disparate genomically.

Psilocybe subaeruginosa CLELAND

CAP: 1–5 cm broad. Conic to obtusely conic, to convex, expanding to broadly convex, often bluntly umbonate. Brown to ochraceous to caramel brown, strongly hygrophanous, fading in drying, lightening to a straw yellow to whitish yellow to dingy grayish white, often tinged with bluing tones. Surface smooth, translucent-striate near the margin when moist. Whitish veil remnants are sometimes present on outer edges of caps when immature. **GILLS:** Attachment adnate to adnexed, ascending, pallid when young, soon darkening with spore maturity to smoky brown to purplish brown to dark chocolate brown with whitish fringes along the margin. **STEM:** 50–125 mm long by 2–5 mm thick, hollow, cartilaginous, thickened and often curved at base, which is typically adorned with thick, radiating rhizomorphs, equal, then narrowing toward the apex. Surface fibrillose, often in patches, whitish to grayish brown, often with bluing streaks, strong bruising bluish where injured. Partial veil richly cortinate, white, fragile, disappearing during maturity or leaving fine fibrillose zones in the upper regions, prone to bluing and can become dusted dark purplish brown with spores. **MICROSCOPIC FEATURES:** Spores purplish brown in deposit, ellipsoid to subellipsoid, 13–15 by 6.6–7.7 µm. Basidia primarily 4-spored, occasionally 2-spored. Pleurocystidia 22–47 by 6–16 µm, fusoid ventricose, mucronate with an elongated neck, narrowing to 2–4.5 µm thick. Cheilocystidia fusoid ventricose, 17–29 by 5.5–11 µm.

HABIT, HABITAT, AND DISTRIBUTION: Solitary to gregarious, sometimes tufted, in soils enriched with woody debris, sandy soils, gardens, wood chips, mixed eucalyptus and pine forests and landscaping wood chips of same. Growing April through November, found primarily in southeastern Australia (with scattered reports from western Australia) and from both the North and South Islands of New Zealand.

COMMENTS: A very potent psilocybin mushroom, high in psilocybin and psilocin, although the published reports likely understate the actual contents due to drying and storage prior to analyses. Strongly bruising bluish. The radiating rhizomorphs at the base of the stem allow this species to be readily transplanted into new wood-chip beds, using the techniques described for *Psilocybe cyanescens* (see page 78). *Psilocybe subaeruginosa* appears to be the center of a constellation of species related to *Psilocybe cyanescens* and *Psilocybe azurescens* and which also may encompass *Psilocybe eucalypta*.

Psilocybe subaeruginosa is very potent and sometimes can cause temporary paralysis in high doses due to unknown constituents. Recovery is usually complete in two days, during which time some report a lack of coordination. Strongly bruises bluish.

Psilocybe subtropicalis GUZMÁN

CAP: 2–3 cm broad. Obtusely conic, expanding to subconvex to subcampanulate to convex, sometimes broadly umbonate, glabrous, translucent-striate toward the margin when moist, reddish brown to brown, hygrophanous, fading to straw color, bruising dark blue to bluish black, darkening in drying especially along the margin. **GILLS:** Adnate or adnexed, pale brown to cinnamon brown or blackish violet, with edges whitish fringed. **STEM:** 40–80 mm long by 1–3 mm thick, even, whitish to concolorous with the cap flesh, hollow, bruising bluish, covered with appressed patches of white fibrils on lower two-thirds and sparsely above, swelling at the base. Partial veil thinly cortinate to absent. Flesh whitish, bruising bluish. **MICROSCOPIC FEATURES:** Spores 5.5–8 by 4–5.5 µm, subrhomboid in face view, subellipsoid in side view, thick-walled, brownish yellow. Basidia 17.5–26.5 by 5.5–8 µm, clavate or subcylindric, hyaline. Pleurocystidia 12–22.5 by 4–7 µm, ventricose, narrowing toward apex, subacute or hyaline, frequent. Cheilocystidia 16–42 by 5–7 µm, abundant ventricose or subcylindric, branching apically, hyaline.

Psilocybe subtropicalis is a terrestrial species growing in subtropical forests of Mexico to Guatemala. It is estimated to be moderately potent.

HABIT, HABITAT, AND DISTRIBUTION: Scattered to gregarious on soils in tropical forests. Reported from Guatemala and eastern Mexico. Likely more broadly distributed than presently reported. This species is infrequently found. It often appears in disturbed soils in subtropical forests.

COMMENTS: Guzmán noted that this species is similar to *Psilocybe herrerae* and also grows in intersections of forest and grassland habitats, where *Psilocybe mexicana* can be found.[76]

Psilocybe tampanensis GUZMÁN & POLLOCK

CAP: 1–2.4 cm broad. Obtusely conic to convex, expanding to broadly convex to plane or umbilicate. Surface smooth, subviscid when moist, soon dry, margin not translucent-striate, lacking a separable gelatinous pellicle. Ochraceous brown to straw brown, hygrophanous, fading in drying to straw yellow to yellowish gray, except for the center, which is often darkened. Flesh often with bluish undertones. **GILLS:** Attachment ascending, adnexed, pallid when young, soon brownish to dark violet brown when fully mature, retaining dull whitish edges. **STEM:** 20–60 mm long by 1–2 mm thick. Equal, enlarging toward the base, adorned with fine patches of fibrils near the apex and tufted with whitish mycelium at base, which is often bluish toned soon after harvesting. Yellowish brown to reddish brown. Flesh whitish to yellowish, bruising bluish where injured. Partial veil finely cortinate, delicate, soon disappearing and typically not leaving an annular zone. **MICROSCOPIC FEATURES:** Spores purplish brown in deposit, subellipsoid to subrhomboid, 8–12 by 6–8.8 µm. Pleurocystidia absent. Cheilocystidia 16–22 by 4–9 µm, lageniform, with a flexuous, extended neck 2–3 µm thick, infrequently, irregularly branching.

Psilocybe tampanensis is a moderately potent grassland species that typically does not show a strong bluing reaction when bruised.

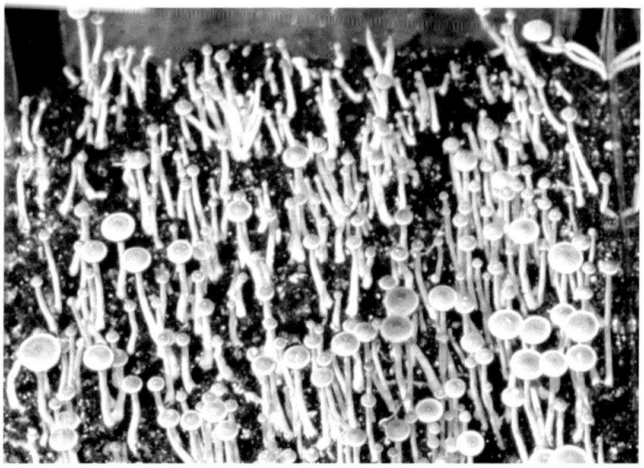

Psilocybe tampanensis mycelium produces sclerotia, which when encased with soil can produce many small mushrooms.

HABIT, HABITAT, AND DISTRIBUTION: Solitary to scattered, in grassy, sandy soils, first found in a sand dune near Tampa, Florida, and later in sandy soil within a deciduous forest in Pearl River County, Mississippi. Additional reports by myconauts from Louisiana, Alabama, and South Carolina are documented on iNaturalist, although these listings have not yet been verified.

COMMENTS: This is a small, nondescript Psilocybe, growing solitary to scattered in sandy, grassy soils. Cultivated fruitbodies can be prolific, and analysis shows moderate potency with 0.68 percent psilocybin and 0.32 percent psilocin.[77] The sclerotia, which are generally less potent than the fruitbodies, have less, 0.19 percent to 0.03 percent, respectively.[78]

The sclerotia of *Psilocybe tampanensis* have been widely cultivated for several decades, utilizing the methods developed by Steven Pollock, MD, who first collected it with Gary Lincoff near Tampa. Methods for its cultivation were further refined and described in *The Mushroom Cultivator*, which I coauthored. This species is in commerce in the Netherlands, where buyers take advantage of a loophole in the law that banned psilocybin mushroom fruitbodies but not the sclerotia, since technically sclerotia are not the same. The sclerotia are hardened masses of mycelial tissue forming in darkness, distinct from the phototropic mushroom fruitbody and are thought to confer an advantage to the species for surviving fire and drought. Mushrooms can emerge, sometimes years later, from the hardened sclerotia under favorable conditions. *Psilocybe mexicana* is a similar grassland species that also produces sclerotia. Both species produce sclerotia prolifically on rye grass seed, millet, and rice stored in the dark for several weeks. These two taxa may be conspecific.

The wide range in potency is attributable to the substrates on which *Psilocybe tampanensis* grows, environmental conditions, methods of drying, storage, and span of time from harvesting to analysis. Sclerotia formation is inhibited by light, whereas light stimulates fruitbody formation directly from the sclerotia. Dr. Gastón Guzmán recollected this species in southern Mississippi in 1996. Since then, numerous reports from iNaturalist show that this species has a wide range of distribution across the southeastern United States. To the untrained observer, *Psilocybe tampanensis* could be mistakenly identified as a grassland *Panaeolus*. See also *Panaeolus rickenii* (= *Panaeolus acuminatus*), *Panaeolus foenisecii*, and *Panaeolus olivaceus* (= *Panaeolus castaneifolius*).

Psilocybe tasmaniana GUZMÁN & WATLING

CAP: 1–1.3 cm broad. Obtusely conic to convex to broadly convex to subcampanulate. Chestnut brown, lightening with maturity to tawny brown or orangish brown, strongly hygrophanous, and in drying becoming dull yellowish to dingy yellow. Surface smooth, viscid when wet, deeply translucent-striate from the marginal two-thirds toward the disc. Edge often decorated with whitish veil remnants when very young. **GILLS:** Attachment adnate, becoming purplish brown as spores mature, with whitish edges that often are bluish tinged. **STEM:** 25–46 mm long by 1–2.5 mm thick. Whitish, often covered with silky fibrils. Bruising bluish, often faint, especially toward stem base. Partial veil finely cortinate, soon disappearing as the cap expands. **MICROSCOPIC FEATURES:** Spores dark purplish brown in deposit, ellipsoid to subellipsoid in side view and subovoid in face view, 10–17 by 7–9 µm. Pleurocystidia few, 17.5–30 by 4.6–10 µm, fusoid-ventricose narrowing to a short apex, 1.6–2.7 µm broad. Cheilocystidia fusoid-ventricose to sublageniform, with a protruding narrowing neck, often two- or three-forked at end, transparent, 22.5–36(44) by 5–10 µm.

HABIT, HABITAT, AND DISTRIBUTION: Solitary to gregarious, in mixed grasslands enriched with dung and wood debris (*Eucalyptus*), from April to December from Tasmania, New Zealand, and Australia.

Psilocybe tasmaniana is a common species in New Zealand, Tasmania, and Australia. There are several forms being called "tasmaniana," and one whose members have distinct nipples may be a new species.

Psilocybe tasmaniana typically grows in clusters or in groups.

COMMENTS: Very often confused with and may be synonymous with *Psilocybe alutacea*, which was published in 2006 by Y. S. Chang and A. Mills.[79]

Psilocybe tasmaniana was published in 1978 by Guzmán and Watling and hence takes priority.[80] This species needs much more analyses to disambiguate the many phenotypes attributed to this taxon, growing in different habitats.

Psilocybe weilii GUZMÁN, STAMETS & F. TAPIA

Common Name: Weil's Psilocybe

CAP: 2–6 cm broad. Campanulate to obtusely conic, with an inrolled margin when young, then incurved, irregular margin, expanding to convex, then broadly convex to nearly plane and uplifting in age. Dark chestnut brown to deep olivaceous brown, typically with a blackish to dark brown zone toward the center. Flesh is 1–5 mm thick at the disc, whitish, bruising bluish. Strongly hygrophanous, fading in drying to pallid to light brown. Surface viscid when moist from a sometimes-separable gelatinous pellicle, translucent-striate along the margin. **GILLS:** Attachment adnate to sinuate with two tiers of intermediate gills, close, even, broad, light brown overall with pallid, floccose edges, becoming dark chocolate brown at maturity. **STEM:** 25–70 mm long by 4–8 mm thick. Equal, enlarging toward the base, protruding with white rhizomorphs. White to dingy brown, bruising bluish overall in age or from drying, covered with sheaths of whitish fibrillose patches below, and pruinose in the superior regions. Cartilaginous, hollow, stuffed with a whitish pith. Partial veil whitish, cortinate, leaving a fibrillose annular zone, often dusted with purplish brown spores. **MICROSCOPIC FEATURES:** Spores dark violet brown to grayish black in deposit, subellipsoid, subrhomboid to subellipsoid in face view, 5.5–6.5 by 4–5 µm in side view. Pleurocystidia abundant, subfusoid or ventricose-rostrate with short apex or sublageniform, 10.5–21.5(24) by 5–9.5(10.5) µm. Cheilocystidia lageniform with single or branched apices, 20–37.5 by 5–6.5 µm.

Psilocybe weilii is a potent psilocybin mushroom and very similar to *Psilocybe caerulescens.* Some experts think they are conspecific.

Psilocybe weilii was first found in pine wood chips used for landscaping near Atlanta, Georgia. It is known from numerous locations in the southeastern United States.

HABIT, HABITAT, AND DISTRIBUTION: Gregarious to clustered, occasionally dispersed, in clay and wood-enriched soils, underneath loblolly pines (*Pinus taeda*) and sweetgum (*Liquidambar styraciflua*) and growing in disturbed grounds such as landscaped areas underneath mixed trees in parks and suburban areas. Fruiting from early September through November in northern Georgia, likely to be more widely distributed than presently reported.

COMMENTS: *Psilocybe weilii* is very close to but genetically distinct from *Psilocybe caerulescens*, according to recent phylogenetic analysis.[81] Both species are terrestrial, growing in pine debris. *Psilocybe weilii* is potent, containing 0.61 percent psilocybin, 0.27 percent psilocin, and 0.05 percent baeocystin.[82] This species is named to honor the lifetime work of Andrew Weil, MD, who for decades courageously advocated the study of psilocybin mushrooms and other psychoactive substances, despite facing fierce criticism from opponents.

Psilocybe weraroa grows on forest debris and in wood-enriched soils.

Psilocybe weraroa BOROVIČKA, OBORNÍK & NORDELOOS

CAP: 3–5 cm long by 1.5–3 cm wide. Roundish, ovoid, or elongated, narrowing toward the base and swelling midway. Whitish to pale bluish gray to olive green, bruising bluish, drying to yellow to dingy brown. Cap is in the form of a sac encasing the gills, the margin incurved, folded, extending down the stem. Surface smooth to tacky. **GILLS:** Tightly arranged spore-producing plates are sepia brown to chocolate brown at spore maturity, coarsely shaped, often elongated, wavy, sometimes chambered. **STEM:** 10–40 mm long by 6 mm thick, slender, equal, whitish to gray, bruising blue or greenish with damage, yellowish brown at the base, covered with fibrils at first, soon becoming smooth in maturity, cartilaginous in age, swelling and curved toward the base, hollow, flesh yellowish orange, thickening in ascent. **MICROSCOPIC FEATURES:** Spores sepia brown to purple brown, 11–17 by 5–8 µm, smooth, ovoid

A secotioid member of the genus, *Psilocybe weraroa* is a mildly potent mushroom that grows in New Zealand. Other secotioid *Psilocybe* species are being discovered elsewhere. See pages 169, 170, and 226. Some mycologists theorize that these secotioid mushroom are mimicking fruit to attract birds who then carry spore populations to distinct locations, since their spores cannot be easily ejected into the air since the gills are encased.

Psilocybe weraroa is at the center of a unique lineage of sac-like psilocybin mushrooms. Note the faint bluish tones due to its psilocin content.

ellipsoid to subellipsoid in shape, rounded at one end with a thin germ pore.

HABIT, HABITAT, AND DISTRIBUTION: Solitary to gregarious, often buried beneath forest litter or māhoe or whiteywood (*Melicytus ramiflorus*), a small lowland tree that often spontaneously colonizes habitats recovering from previous disturbances from logging or other destructive practices. Known only from New Zealand. Other habitats include a close association with mulch from the widely planted *Cordyline australis* (which Māori call *tī kōuka*) cabbage trees that are being widely planted in parks, gardens, and along streets in New Zealand. In more forested habitats, *Psilocybe weraroa* also appears in the detritus below overhanging forest ferns (*Cyathea* spp.). Sometimes plentiful in lowland mixed rainforest near Auckland and Wellington. Difficult to find, as it is often hidden underneath the top layers of decomposing debris.

COMMENTS: One of a number of curiously different, recently described psilocybin species, morphologically unique compared to the vast majority of other species in the genus *Psilocybe*. Phylogenetic analysis places this psilocybin-active species closest to the *Psilocybe cyanescens* complex. *Psilocybe weraroa* is a secotioid fungus, an intermediate stage between the evolution of gasteroid (puffballs, truffles) and gilled mushrooms like the classic agarics and typical *Psilocybe* species.

The chambered spore-producing layers are encased in a sac and hence not readily airborne. Some mycologists speculate that these secotioid species are evolving gills, but not yet developed a form to release airborne spores from engulfed gill plates. Insects and birds are likely vectors of spore dispersal. This species has reportedly caused the wood lover's paralysis (see page 142) that is also reported from *Psilocybe azurescens*,

Psilocybe cyanescens, and *Psilocybe subaeruginosa*.[83] Another curious species that has immature gills embedded and is psilocybin-active but not yet named has been found near Lake Tahoe, possibly belonging to the genus *Galeropsis* (see page 227). See also *Psilocybe maluti*, a grassland species from Africa, which is also secotioid. I expect we will find more of these secotioid and gastroid-like species.

Often hidden from view, this bluing species can be confused with another secotioid, the blue pouch fungus, *Clavogaster virescens*, which can also be pale blue but has not yet been reported to contain psilocybin, nor bruise bluish. *Clavogaster virescens* does not have gills, but its spores are produced from a chambered like gleba, whereas *Psilocybe weraroa* still has distinct but often contorted, even chambered gills, encased by a pouch-like fruitbody.

Psilocybe yungensis SINGER & A.H. SMITH = *Psilocybe acutissima* HEIM

Common Names: Hongo Adivinador (Divinatory Mushroom), Hongo Genio (Genius Mushroom)

CAP: 0.5–2.5 cm broad. Acutely conic to conic to campanulate at maturity, often with a sharp umbo. Surface smooth, viscid and translucent-striate most of the way to the disc when moist, pellicle not separable. Rusty brown to orangish brown to dark reddish brown, hygrophanous, fading to dull yellowish brown or dingy straw in drying. Bruising bluish where injured and then often dark blackish with bluish tinges in drying. **GILLS:** Attachment adnate to adnexed, close to crowded, dull gray at first, brown, then purplish brown with spore maturity. Edges whitish to pale. **STEM:** 25–60 mm long by 1.5–3 mm thick, equal to enlarging toward the base. Surface covered with a sheath of dense whitish fibrils, pale brownish above and reddish brown to reddish brown-black near the base. Flesh bruising bluish, hollow, and fairly brittle. Partial veil cortinate, soon disappearing with maturity, leaving whitish fibrils along the cap margin and scant remnants on the upper regions of the stem. **MICROSCOPIC FEATURES:** Spores dark purplish brown in deposit, rhomboid to subrhomboid to subellipsoid, 4.4–7 by 4–6 µm. Pleurocystidia 14–25 by 4.4–10.5 µm, ventricose below and mucronate at the apex. Cheilocystidia 14–33 by 4.4–7.7 µm, variable, ventricose to clavate to strangulated.

Psilocybe yungensis is found across a broad region from Mexico to Brazil to the Caribbean.

Psilocybe yungensis is potently psilocybin-active.

HABIT, HABITAT, AND DISTRIBUTION: Most frequently found in clusters or gregariously on rotting wood, sometimes at the bases of stumps, in coffee plantations and/or subtropical forests, primarily found March to October. Found in northeastern, central, and southeastern Mexico; Bolivia; Colombia; Ecuador; Brazil; and reported from the Caribbean island of Martinique. Likely to be more widely distributed than presently reported.

COMMENTS: A species whose Spanish name means "genius mushroom," historically used by the Mazatec Indigenous people of Mexico. This species' distinctively sharp umbo is one of several features making it readily identifiable. This mushroom is distinct for its penchant for growing in great numbers on decomposing stumps or wood debris, its orangish color, and that the caps remain conic at maturity. See also the more nippled *Psilocybe hoogshagenii* and the much rarer, smaller *Psilocybe stametsii*, the latter of which also grows in Ecuador but has smaller spores.

Psilocybe zapotecorum HEIM emend GUZMÁN

Common Names: Badao Zoo (Zapotec) or Hongo Borracho (Spanish; Drunken Mushroom), Bi-neechi (Zapotec; Crown of Thorns Mushroom)

CAP: 1-3 by 7-11 cm, highly variable in form, conic to convex to campanulate, sometimes subumbonate or bluntly umbonate and convoluted in age. Surface smooth, translucent-striate near the margin when moist. Reddish brown to organic brown, hygrophanous, fading to beige, orangish rose to straw in drying, quickly bruising blue to green to blackish where injured or in age. **GILLS:** Attachment sinuate or adnate, pale brown to purplish brown to violet purple, with edges concolorous with gill face or slightly paler. **STEM:** (40)100–200 mm long by (3)5–10(12) mm thick, equal to slightly expanded at the base, sometimes with a pseudorhiza; at times flexuous, irregular in thickness. White to grayish to variably reddish brown or vinaceous, bluing when touched or injured, with blackish violet tones. Surface covered with whitish fibrillose patches, sometime strongly scabrous-strigose near the base. **MICROSCOPIC FEATURES:** Spores purplish, violet brown in deposit, oblong-ellipsoid, (5.5)6.6–7.0(8.8) by 3.8–4.4(5.5) µm. Pleurocystidia 20–38 by 5.5–14 µm, variable in form, fusoid-clavate, ventricose to submucronate, sometimes with irregularly divided apices. Cheilocystidia 13–27 by 3.5–6 µm, ventricose to fusoid, pyriform to lageniform, with an extended neck 1.5–2.2 µm thick.

HABIT, HABITAT, AND DISTRIBUTION: Cespitose to gregarious, rarely scattered; in swampy or muddy soils, humus rich with leaves and wood debris, sandy, marshy deciduous forests, and coffee

Psilocybe zapotecorum in its pristine form. The mushrooms bruise strongly bluish upon handling or when damaged.

plantations. Frequently found on the faces of ravines with exposed soils. Found in southern Mexico at 2000 feet (600 m) elevation and subtropical South America. Collected in Colombia, Peru, Brazil, Uruguay, and Argentina.

COMMENTS: A potent and strongly bluing mushroom when bruised, *Psilocybe zapotecorum* is comparatively large and can sometimes be covered with sand as it forces its way up through soils. Not surprisingly, Heim and Wasson (1958) reported, based on Albert Hofmann's analysis, 0.05 percent psilocybin and 0 percent psilocin in 2-year-old specimens.[84] One of the most curious species in the genus *Psilocybe*, this mushroom typically has an asymmetrical cap that is often convoluted in form. It is held in high esteem by native Mazatecs and Zapotecs in the state of Oaxaca, Mexico. Guzmán reported in his 1984 monograph that this mushroom is sometimes found inside the mud houses of the Zapotecs, which may be descendent from bringing the mushrooms into their dwellings.[85] This species seems like an excellent candidate for creating naturalized patches using stem-butt inoculations.

The Genus *Panaeolus*

The genus *Panaeolus* is in the Bolbitaceae family and hosts a few psilocybin mushrooms. *Panaeolus* is Greek for "mottled," so named for the distinctively shaded gills with varying regions of basidia that mature to produce black spores that give the gills a variegated or spotted appearance. Mushrooms in this genus have caps whose cuticles are typically made of rounded cells, along with blackish spores (with a few exceptions, such as *Panaeolus foenisecii*, which has dark brown spores). In contrast, the cap cuticles of *Psilocybe* species are composed of interlacing filaments and typically have purplish brown spores.

The faintly bluing *Panaeolus cinctulus* is weakly psilocybin-active. The strongly bluing and potent *Panaeolus cyanescens* is at the center of a constellation of similar-looking species that have also been placed into the genus *Copelandia*. Typically pastoral, the psilocybin-active *Panaeolus* species mostly grow on dung, in well-manured soils and grasslands, and occasionally on composting straw. They typically do not grow on wood debris. Strauss et al. published a good review article of twenty worldwide reports of the reputed psilocybin-active species among the fifty or so nonactive species of *Panaeolus*.[86]

Panaeolus africanus OLA'H

Common Name: Elephant Dung Panaeolus

CAP: 1.5–2.0 cm broad. Obtusely conic, hemispheric and broadly convex in age. Surface smooth, viscid when wet, especially in young specimens, reticulated along the cap margin. Creamy white to grayish, sometimes reddish brown toward the disc, and becoming grayish brown in age. Margin incurved when young, often irregular, and non-translucent. Flesh grayish white, bruising bluish. **GILLS:** Attachment adnate to adnexed, sometimes sinuate, rarely subdecurrent, widely spaced, irregular, grayish at first, soon grayish black, to blackish in age, and mottled as spores mature. **STEM:** 30–50 mm long by 4–6 mm thick, equal, firm, pruinose toward the apex. Whitish to white with pinkish tones, generally lighter than the cap, lacking any veil remnants. **MICROSCOPIC FEATURES:** Spores nearly black in deposit, 11.5–14.5 by 7.9–10 µm, lemon shaped, often variable. Basidia 2- and 4-spored. Pleurocystidia present, with extended sharp apices, 25–50(60) by 10–17(20) µm. Cheilocystidia clavate, 17–24 by 7.4–12 µm.

HABIT, HABITAT, AND DISTRIBUTION: Reported from central Africa to the southern regions of Sudan, Chad, Central African Republic, India, and Mexico. Probably more widely distributed across Central America than currently known. Found on hippopotamus and elephant dung in

Panaeolus africanus is a light-colored psilocybin-active species that bruises bluish and is found on cow, elephant, and hippopotamus dung.

the spring or during the rainy seasons, and in zoos, all of which can become a nexus for further distribution to other regions of the world as animals are dispersed.

COMMENTS: Some collections of *Panaeolus africanus* can be relatively robust for a member of this genus, having a comparatively thick stem and a viscid cap when wet. Contains psilocybin and psilocin, according to Ola'h,[87] although in irregular amounts. Years ago, I found this species on zoo-doo, elephant dung from the Seattle Zoo. I suspect that the importation of exotic animals to zoos and the subsequent selling of their dung to gardeners is one vector by which this species is spreading from the heartland of the African continent to other regions of the world.

Panaeolus cinctulus (BOLTON) SACCARDO = *Panaeolus subbalteatus* BERKELEY & BROOME

Common Names: Haymaker's Panaeolus, Belted Panaeolus

CAP: 2–6 cm broad. Convex to campanulate, then broadly convex, finally expanding to nearly plane with a broad umbo. Cinnamon brown to orangish cinnamon brown, fading to tan in drying with a dark brown encircling zone around the margin. **GILLS:** Attachment adnate to uncinate, close, slightly swollen in the center, with three tiers of

Panaeolus cinctulus is a mildly psilocybin-active species that thrives in horse dung, often found around horse paddocks or racetracks.

Panaeolus cinctulus grows on composting straw. Fruitings are enhanced when enriched with horse dung.

intermediate gills. Color brownish and mottled, blackish when fully mature, often fringed with whitish edges **STEM:** 50–60 mm long by 2–4 mm thick, brittle, hollow, fibrous, breaking apart in longitudinal strips. Reddish beneath minute whitish fibrils, darkening downward. Often bruising bluish at the base, most evidently in the basal mycelium. **MICROSCOPIC FEATURES:** Spores black in deposit, lemon shaped in side view, subellipsoid in face view, 11.5–14.0 by 7.5–9.5 µm. Basidia 2- and 4-spored. Pleurocystidia absent. Cheilocystidia variable in form, mostly pear shaped, 14–21 by 3–7 µm.

HABIT, HABITAT, AND DISTRIBUTION: Growing cespitosely to gregariously in dung or in well-manured ground in the spring, summer, and early fall. Sometimes found on aging composting bales of straw. Widely distributed across the world. Reported from North America, South America, Europe, Russia, Asia, Africa, Australia, New Zealand, and the Hawaiian archipelago. Like *Panaeolus cyanescens*, this is one of the most widely distributed psilocybin mushrooms in the world.

COMMENTS: Also commonly listed in field guides as *Panaeolus subbalteatus, Panaeolus cinctulus* is

Panaeolus cinctulus and many *Panaeolus* species have mottled gills because basidia mature with spores regionally.

weakly to moderately psilocybin-active, featuring a distinctive dark band ringing the outer edges of the cap, expanding to nearly plane in age, with a bluing reaction most commonly seen at the stem base and in the basal mycelium. Stijve and Kuyper found maxima of 0.14 percent psilocybin, 0 percent psilocin, and 0.033 percent baeocystin.[88] From Brazilian specimens, Stijve et al. found 0.08 percent psilocybin and no detectable psilocin. Jochen Gartz reported maxima of 0.7 percent psilocybin, 0 percent psilocin, and 0.46 percent baeocystin.[89] Gurevich found 0.14 to 0.36 percent psilocybin from specimens collected in central Russia and middle Siberia.[90] These analyses are dated, however, and I suspect that they understate the actual content in fresh, air-dried specimens.

I most often find *Panaeolus cinctulus* directly on horse dung or composts made with horse manure. This species is widely cultivated, often unintentionally, on the discarded leavings (manure and straw) from mucking out horse stables. The horse manure and straw from race-tracks often result in bountiful fruitings of this species, especially in late spring. This species also comes up in the composts used by farms growing button mushrooms (*Agaricus*), especially when horse manure is a component. It also grows directly out of composting haybales, hence the name Haymaker's Panaeolus, which is also the common name for the psilocybin-inactive *Panaeolus foenisecii.*

When I have consumed this mushroom raw, it made me queasy, a symptom that might have been alleviated had I made a tea. *Panaeolus cinctulus* seems to harbor a lot of other organisms—and is a favorite home for insect larvae. It is highly perishable and should ideally be picked young, before full spore maturity, and dried quickly. Once harvested, the mushrooms quickly become limp, whereas the *Psilocybe* species tend to have a more stable structure due to their more cartilaginous nature.

Another Panaeolus that can develop a dark band along the margin, especially as it dries in situ, is *Panaeolus rickenii* (= *Panaeolus acuminatus*); see page 126. However, the consistent dark band around the flaring cap margin of *Panaeolus cinctulus* and the bluing reaction in the basal mycelium of the stem are the most prominent macroscopic feature distinguishing this species from *Panaeolus rickenii* and other close look-alikes.

Panaeolus cyanescens SACCARDO = *Copelandia cyanescens* (BERKELEY & BROOME) SACCARDO = *Copelandia papilionacea* (BULLIARD EX FRIES) BRESADOLA

CAP: 1.5–3.5(4) cm broad. Hemispheric to campanulate to convex at maturity. Margin initially translucent-striate when wet, incurved only in young fruitbodies, soon opaque and decurved, expanding in age, becoming flattened and often split or irregular at maturity. Light brown at

Panaeolus cyanescens is a potent, dung-dwelling, strongly bluing species that is often whitish in color at first. Sunlight exposure often lightens the color of the caps. Several closely related species resemble this iconic species, all of which are psilocybin-active and bruise strongly bluish.

Panaeolus cyanescens can be tawny in the shade, whitening when exposed to sun, but darkening in age from spore maturity.

Sometimes *Psilocybe cubensis* and *Panaeolus cyanescens* can co-occur in the same pasture, sometimes upon the same pile of dung.

Panaeolus cyanescens soon bruises bluish from handling or with age.

Panaeolus cyanescens is at the center of a clade of closely related psilocybin-active Panaeoli, including *Panaeolus chlorocystis,* featured here.

first, beige to pallid gray or nearly white overall with the center regions often remaining tawny brown, soon fading. Cap cracking horizontally in age with irregular fractures and in drying. Flesh readily bruising bluish. **GILLS:** Attachment adnexed, close, thin, with two or three tiers of intermediate gills. Mottled grayish black at maturity. **STEM:** 65–115 mm long by 1.5–3.0 mm thick. Equal to bulbous at the base, tubular. Often grayish toward the apex, pale yellowish overall, then flesh colored to light brown toward the base; readily turning bluish when bruised. Surface covered with fine fibrillose flecks that soon disappear. Partial veil absent. **MICROSCOPIC FEATURES:** Spores black in deposit, 12–14 by 7.5–11.0 µm, nontransparent, and without granulations. Basidia 4-spored, occasionally 2-spored. Pleurocystidia fusoid-ventricose, narrowing to an acute apex, 30–60(80) by 12–17(25) µm. Cheilocystidia present, 11–15 by 3–5(6) µm.

HABIT, HABITAT, AND DISTRIBUTION: Growing scattered to gregariously on dung, in pastures and fields in the equatorial and subtropical regions of the world. *Panaeolus cyanescens* is one of, if not *the*, most widely distributed psilocybin-active mushroom in the world. Widespread in most semitropical zones and reported from the southern United States (Hawaii to California to Texas to Tennessee, Kentucky, North Carolina, Virginia) to Mexico, across Central and South America, Asia, Australia, and in warmer parts of the Mediterranean region.

COMMENTS: This species is a strong producer of psilocybin and/or psilocin, although wild specimens are highly variable in content. Two notable reports show 1 percent and 2.5 percent combined psilocybin and psilocin.[91,92] I think exposure to sunlight (UV) degrades the psilocybin content in these small, fragile mushrooms. *Panaeolus cyanescens* is at the center of a constellation of other potent, psilocybin-active species that are very similar in appearance and also dung dwelling. The constellation of bluing species that are taxonomically interrelated to *Panaeolus cyanescens* are *Panaeolus africanus, Panaeolus bisporus, Panaeolus chlorocystis, Panaeolus cambodigeniensis, Panaeolus microsporus*, and *Panaeolus tropicalis*, some of which are considered synonyms. All can quickly become ridden with maggots and small bugs, especially in age.

Once I was called out to a house near Tenino, Washington, to identify an unknown mushroom. It turned out to be *Panaeolus bisporus* (notably it has 2-spored, not 4-spored, basidia). The homeowner had recently brought in manure from a local horse stable (that had recently received horses from Florida) to fertilize his yard. The mushrooms showed for 2 years and then petered out. This species is easy to grow in a lab; for information on lab cultivation, see *The Mushroom Cultivator*, a book I coauthored with Jeff Chilton.

Panaeolus olivaceus F.H. MØLLER = *Panaeolus castaneifolius* (MURRILL) OLA'H

CAP: 1.0–3.0(4) cm broad. Distinctly campanulate at first, soon subhemispheric, then convex and becoming broadly convex in age. Margin incurved at first, soon straightening, not appendiculate, and slightly striated. Dark smoky gray when moist, hygrophanous, soon drying to a more straw yellow or pale ochraceous and remaining more reddish brown at the apex and smokier brownish along the margin. Surface sometimes finely wrinkled. **GILLS:** Attachment adnate to adnexed, close, and thin. Pallid at first, becoming dark purplish gray black at spore maturity. **STEM:** 40–60(75) mm long by 3–4(6) mm thick. Equal, narrowing toward the base. Hollow or tubular,

The black-spored *Panaeolus olivaceus* (= *Panaeolus castaneifolius*) is close to the brown-spored *Panaeolus foenisecii*, with which it is often confused.

brittle. Grayish to ochraceous or tan at the base. Surface slightly striated, pruinose. **MICROSCOPIC FEATURES:** Spores black in deposit, finely roughened, 12–15 by 7–9.5 µm. Pleurocystidia few or absent, not projecting beyond the plane of the basidia. Cheilocystidia 20–28(35) by 7–10 µm.

HABIT, HABITAT, AND DISTRIBUTION: Growing scattered to gregariously in grassy areas across North America, northern South America, Europe, and Russia. Probably more widely distributed than currently known.

Panaeolus olivaceus is weakly psilocybin-active and common in grassy habitats.

COMMENTS: Originally described by Murrill in 1923 as *Psilocybe castaneifolia*; in 1948 Alexander Smith transferred it to the genus *Panaeolus*, renaming it *Panaeolus castaneifolius*.[93,94] *Panaeolus olivaceus* is weakly psilocybin-active, if at all. Some collections of this species do not contain psilocybin according to a recent analysis.[95] Distinguished from *Panaeolus foenisecii* by the color of the mature gills and more definitively by the color of the spore deposit, which are very dark purplish gray black in *Panaeolus olivaceus* but brown in *Panaeolus foenisecii*. (See also *Panaeolus rickenii* on page 126.) Some collections of *Panaeolus olivaceus* show ephemeral bluing in the basal mycelium but rarely in the fruitbodies. Notably, more studies are needed on the varieties within this species. Past analyses may have been based on misidentified specimens as this species and *Panaeolus cinctulus* are commonly confused for one another. Until more studies are conducted with accurately identified specimens, it is questionable that this species produces psilocybin and/or psilocin. It is, however, a good indicator species, co-occurring with many psilocybin-active species.

The Genera *Conocybe, Conocybula,* and *Pholiotina*

The genus *Conocybula* is a spinoff from the genus *Pholiotina* and ultimately *Conocybe*. Both are classic little brown, typically conic shaped, comparatively long-stemmed mushrooms with rusty brown spores. Several former *Conocybe* species are now called *Pholiotina*, a genus first named by V. F. Fayod in 1889 for *Conocybe*-like mushrooms with partial veils, but the genus has now been expanded to include those without partial veils.[96] *Conocybe* means "cone head" whereas *Pholiota* means "mushrooms with scales," and *Pholiotina* is a diminutive form of the word, referencing their smaller sizes of species in this genus. Similarly, *Conocybula* is the diminutive term for *Conocybe.* Pholiotinas have cystidia that are subcapitate to blunt, whereas *Conocybe* cheilocystidia are lecythiform (shaped like bowling pins). Most recently, the genus *Conocybula,* formerly a subgenus within *Pholiotina,* was elevated to the genus level to encompass the formerly bluing, psilocybin-active Pholiotinas based on phylogenetic analsyses. These are LBMs, little brown mushrooms, which have traditionally been challenging for many conventional mycologists to decipher.

Notably, there are both psilocybin-active and toxic mushroom species in the genus *Pholiotina,* which increases the risks of misidentification for those seeking psilocybin mushrooms.[97,98]

Typically, *Conocybe, Conocybula,* and *Pholiotina* species grow scattered to gregariously and have thin, long, fragile stems and conic caps. The habitats in which these genera grow vary from dung and grass to decayed wood.

Two species of the *Conocybula* are notable: *Conocybula cyanopus* and *Conocybula smithii*, which may eventually be proven to be synonymous. Other species that are probably psilocybin-active include *Conocybe siligineoides*, which was first found in Mexico by Heim and Wasson and reported to be used for shamanic purposes by the Mazatecs of Oaxaca (no analyses or confirmation of the properties of this species has been reported since its discovery circa 1956). Another reportedly psilocybin-active species from Finland is *Conocybe kuehneriana,* a putative synonym of *Conocybe velutipes.*[99] Both these species currently remain in the genus *Conocybe.* With time, some of the above-named species can bruise bluish, although sometimes not markedly so—or not at all—and usually just at the base of the stem. Both species are very small and lack a ring entirely. But, in view of the existence of the deadly *Pholiotina rugosa* (= *Pholiotina filaris*) and its not-always-present annulus and often darkening at the stem base, I strongly caution against amateurs experimenting with species of *Conocybe* and *Pholiotina.* Moreover, Dr. Michael Beug notes that there are at least five other potentially deadly sister species similar to *Pholiotina rugosa*, currently unnamed, growing in the Pacific Northwest of North America. There is an abundance of other dark purplish brown–spored *Psilocybe* species and

Panaeolus species that do not present the dangers inherent within these rusty brown–spored genera. Proceed with the utmost caution! Best to observe, not to ingest.

Conocybula cyanopus (G.F. ATK.) T. BAU & H.B. SONG
= *Conocybe cyanopus* (ATKINS) KUHNER
= ? *Pholiotina smithii* (WATLING) ENDERLE
= *Conocybula smithii* (WATLING) T. BAU & H.B. SONG

Common Name: Blue Footed Cone Head

CAP: 0.7–1.2(2.5) cm broad. Bluntly conic, expanding in age to nearly hemispheric to convex, expanding to broadly convex with age. Margin translucent-striate when moist and sometimes appendiculate at first with fine fibrillose remnants of the partial veil. Rusty to reddish cinnamon brown to dark brown. Surface moist when wet, soon dry, smooth overall to slightly wrinkled toward the disc with age. **GILLS:** Attachment adnexed, close, moderately broad. Dull rusty brown with a whitish fringe along the margin. **STEM:** 20–50 mm long by 1–1.5 mm thick. Equal to slightly curved at the base, fragile, easily breaking. Whitish at first, becoming grayish or brownish at the apex, sometimes bluing, and often adorned with whitish mycelium at the base that has bluish overtones. Partial veil thinly cortinate, sometimes leaving trace remnants along the cap margin, soon disappearing. Typically, without an annular zone or annulus. **MICROSCOPIC FEATURES:** Spores rusty brown in deposit, 6–9 by 4–5 µm. Pleurocystidia absent. Cheilocystidia 14–30 by 4–12 µm.

Conocybula cyanopus, the Blue-Footed "Conocybe" Cone Head, is strikingly beautiful although extremely fragile. Note bluing reactions at the stem bases. These specimens are atypically large.

HABIT, HABITAT, AND DISTRIBUTION: Scattered in grassy areas, mossy lawns, and fields in the summer and fall. Reported from Oregon, Washington, and Colorado to Virginia, Pennsylvania, New York, and Maine. Also found in Vancouver and Quebec, Canada, in Europe, China, the Ukraine, and one report from Russia (Moscow). *Conocybula cyanopus* is probably widely distributed across the temperate regions of the northern hemisphere but goes unnoticed because of its minute stature.

COMMENTS: Appropriately named, *cyanopus* means "blue foot." Infrequently found, often solitary or scattered in grassy-mossy habitats, and usually not in large quantities, this petite mushroom is very fragile. Pulling the mushroom up usually breaks off the base of the stem, which is often attached to dead grass. If the stem base is broken off, an essential identification feature is lost: the characteristic bluing, complicating macroscopic identification. Potently active although petite in size, Michael Beug and Jeremy Bigwood found 0.93 percent psilocybin but no

Conocybula smithii and *Conocybula cyanopus* both bruise bluish, typically at the stem bases. These two species are so closely related that they may be one species. Both species are weakly psilocybin-active, rarely found in sufficient quantities to attract the attention of those seeking high-dose psilocybin journeys. *Conocybula smithii* reportedly has a faint fuzz on the surface of the caps because of elongated pileocystidia and has serrated gill edges, two features of questionable taxonomic significance.

psilocin.[100] A. L. Christiansen et al. reported ranges of 0.33–0.55 percent psilocybin and 0.004–0.007 percent psilocin.[101] Jochen Gartz found 0.78–1.01 percent psilocybin, 0 percent psilocin, and 0.12–0.20 percent baeocystin.[102] M. Halama et al. reported 0.90±0.08 percent psilocybin, 0.17±0.01 percent psilocin, 0.16±0.01 percent baeocystin, 0.053±0.004 percent norbaeocystin, and 0.011±0.0007 percent aeruginascin.[103]

Conocybula smithii is virtually identical to *Conocybula cyanopus*. Hamala et al. reported one collection with up to 1 percent psilocybin, but most collections had only a small fraction of that. They further noted that "*Pholiotina smithii* is virtually identical except that it has more cinnamon-brown lamellae, more distinctly striate and paler pileus, and somewhat wider cheilocystidia." Together, the suggested distinctions between these two taxa are weak. DNA studies are needed to compare the type collection of *Pholiotina smithii*, but the type collection consists of one specimen. Hausknecht et al. noted "The specimen designated as 'part of neotype' consists of one strongly collapsed fruitbody, which has almost no intact lamellae."[104]

See also *Pholiotina aeruginosa*, *Pholiotina atrocyanea*, and *Pholiotina sulcatipes*, which all can have bluish tones. Look-alikes that are not bluing include *Conocybe velutipes*, *Conocybe tenera*, and the deadly *Pholiotina rugosa* (= *Pholiotina filaris*). Confusing *Pholiotina rugosa* (= *Pholiotina filaris*) with these bluing *Conocybula* species can have deadly consequences (see page 123 for a photograph of *Pholiotina rugosa*).

The Genus *Gymnopilus*

Of the more than 200 species in the genus *Gymnopilus*, several are psilocybin-active species and a few are poisonous. The vast majority of Gymnopili have not yet been analyzed. Many new psilocybin-active species have yet to be published. These mushrooms have complex chemistries, confused taxonomic histories, and sparse records of safe historical usage despite their large size and bright colorations.[105]

Gymnopilus means "naked cap" or "smooth cap." These medium to large mushrooms are wood lovers with caps whose surfaces are dry; many have well-developed partial veils, with most leaving a membranous annulus or distinct annular zone often dusted with rusty brown spores. Many of these mushrooms are an overall orangish brown in color. A few bruise bluish and are psilocybin-active.

I highlight five species that bruise bluish: *Gymnopilus dilepis, Gymnopilus luteofolius, Gymnopilus luteus, Gymnopilus purpuratus*, and *Gymnopilus subspectabilis*. There are very few bioassays by people who have ingested these mushrooms to testify to the quality of the experiences of psilocybin-active *Gymnopilus* species, except for sometimes amusing accidental ingestions.[106] In my opinion, it is better to know these species than to consume them. However, I am fascinated by the complex biochemistry of these mushrooms, especially in the potential neuroregenerative and, in the case of the *Gymnopilus junonius* cluster, the neurotoxic properties yet to be explored. (Some mushroom toxins prove to be medicines at lower doses, like illudins M & S in *Omphalotus illudens*, a close relative of *Omphalotus olearius*, for killing conventional medicine–resistant tumor cell lines.[107]) Some of the other *Gymnopilus* species known to be psilocybin-active are *Gymnopilus braendlei, Gymnopilus luteoviridis*, and possibly *Gymnopilus validipes*, yet another species in this genus that has a confused taxonomic history.

At least one species, identified as *Gymnopilus junonius* (= *Gymnopilus spectabilis* s. auct.), is known to contain neurotoxic oligoisoprenoids and thought by some to contain psilocybin, but analyses may have confused two closely allied species.[108] *Gymnopilus spectabilis* has historically been listed as containing psilocybin and psilocin, but the actual specimens tested also have

Gymnopilus ventricosus, the Western Jumbo Gymnopilus, is a psilocybin-inactive species that can become very large. Many species are brightly colored and majestically beautiful. Some species closely related to *Gymnopilus ventricosus* may contain psilocybin, like *Gymnopilus voitkii*.

a confused taxonomic history and were likely misidentified.[109,110,111] *Gymnopilus junonius* has not yet been found in North America but can be found in Europe, South America, and Australia.[112] Species closely related to *Gymnopilus junonius* have been found in North America but their chemistries remain unstudied. I do not recommend consuming *Gymnopilus* species because of their complex chemistries and potential for misidentification with the toxic *Gymnopilus* species, *Omphalotus illudens* (Jack O'Lantern), and the smaller but deadly *Galerina marginata* (Autumnal Galerina), all of which can have orangish brown caps and similar spore colors. Given that there are more than 200 species in this genus, I expect more psilocybin-active and poisonous species are likely to be discovered. Be forewarned. Be careful. *Gymnopilus* species are not recommended for consumption. Very few people have eaten these repeatedly, and their safety is uncertain. In my opinion, it is ill-advised to be the first ones to ingest, or bioassay, these species.

Gymnopilus luteofolius (PECK) SINGER = ? *Gymnopilus aeruginosus* (PECK) SINGER

Common Name: Yellow-Gilled Gymnopilus

CAP: 2–8 cm broad. Convex at first, expanding with maturity to broadly convex to nearly plane. Surface dry, covered with dense, often fibrillose, appressed purple red to orangish scales, disappearing in age. Dark red to reddish brown, fading in drying to pinkish red or yellowish red, and eventually yellow; sometimes bruising bluish green. Margin even, inrolled to incurved when young, soon straightening, adorned with fibrillose veil remnants. Flesh thick, reddish to purplish, then fading to yellowish in drying. **GILLS:** Attachment adnate to sinuate, to slightly

Gymnopilus luteofolius is known for its magenta cap and bright yellow gills, bruising bluish, at times evident at the base of the stems.

decurrent, close to subdistant, broad, initially yellow, then rusty orange, with serrated edges. **STEM:** 30–80 mm long by 3–10 mm thick, equal to enlarged downward when forming singly, narrowing when forming in clusters, often curved at the base. Yellowish or stained rust-colored in age. Base of stem sometimes bruising bluish. Partial veil densely cortinate to nearly membranous, leaving a superior membranous annulus or fibrillose annular zone soon dusted with rusty orange spores. **MICROSCOPIC FEATURES:** Spores rusty orange in deposit, ellipsoid, roughened, 5.5–8.5 by 3.5–4.5 µm. Pleurocystidia 30–38 by 5–10 µm, fusoid to subventricose. Cheilocystidia 23–28 by 4–7 µm, ventricose to flask shaped with or without a swollen head.

HABIT, HABITAT, AND DISTRIBUTION: Commonly found on hardwood debris made from oak, alder, or eucalyptus, especially for the first 2 years, and on wood chips used in nurseries and landscaping, from California to British Columbia from June through November. Found also in Texas, Florida, New York, Tennessee, Michigan, and New Mexico. Hesler reported that this species is widely scattered across the United States on conifer wood,[113] which has not been true from my limited experience. Also occurring on stumps

Gymnopilus luteofolius from British Columbia, Canada. Note the faint bluish tones on the cap margins.

and logs. Curiously, I have found this mushroom sprouting from the alder boards of my wooden boat a week or two after washing it down. And a friend brought me specimens fruiting from the hardwood (alder) trim of another boat a few miles away. The mycelium of this mushroom seems to incubate within alder and other trees. Since alder is commonly used in boats and in making furniture, this species is unwittingly spread by humans.

COMMENTS: The epithet *luteofolius* means "yellow." Mildly active, *Gymnopilus luteofolius* bruises bluish, especially in cold weather. I have collected this mushroom on oak chips (*Quercus lobata* and other species) near Sebastopol, California, and on alder (*Alnus rubra*) chips on Cortes Island, British Columbia. The phenotypes of the California collections I have seen have brighter yellow gills than those from British Columbia. Clones of collections produce mushrooms in culture on nutrient-filled Petri dishes, a strong indication that this species can be grown on sterilized media. I suspect this mushroom can easily be cultivated on wood substrates outdoors in beds of wood chips. See also *Gymnopilus dilepis*, which is very similar.

Gymnopilus luteus (PECK) HESLER

Common Name: Yellow Gymnopilus

CAP: 3–20 cm broad. Hemispheric, then convex to broadly convex, expanding during maturity to broadly convex, with an irregularly infolded, overhanging margin. Yellowish to yellowish orange to warm buff orange, darkening toward the center; dry, smooth, covered with finely floccose fibrils, silky, and at times decorated with floccose-squamulose fibrillose scales toward the center. Flesh pale yellow to orangish brown and sometimes bruising bluish green. **GILLS:** Attached, adnexed, thin, close, pale yellow, becoming rusty brown with age and spore maturity. **STEM:** 40–100 mm long by 5–30 mm thick, equal to slightly enlarging toward the base; firm, solid, concolorous with the cap, bruising yellowish rusty stains, sometimes with bluish green tones, naturally or when handled; finely fibrillose, longitudinally strigose. Partial veil thinly membranous, fugacious, usually forming fragile temporary membranous, torn annulus but more typically annular zone in the upper regions, dusted orangish brown from spores. Overall color orangish brown, occasionally bruising bluish green or naturally in age. **MICROSCOPIC FEATURES:** Spores rusty brown orange in deposit, 6.5–8.3 by 4.5–5.7 µm, ellipsoid to dextrinoid, surface finely roughened to warty, lacking a germ pore. Pleurocystidia rare or absent. Cheilocystidia lageniform to lecythiform but sometimes without swollen apices, 19.3–35.4 µm. Clamp connections present.

HABIT, HABITAT, AND DISTRIBUTION: Widely distributed across the eastern United States, Canada, Europe, and Russia; with a few reports from Thailand, southern India, and eastern Australia. This species is usually found growing solitary

Gymnopilus luteus is a bright yellow species whose bluing reaction is not always obvious. As specimens age or are damaged, streaks of blue tones can often be seen as is evident in the small specimen in lower left of cluster. Pristine specimens may show no bluing reaction at all, until hours after they have been harvested and from handling.

or in small groups from June to September on decomposing hardwoods, sometimes conifers.

COMMENTS: Weakly to moderately psilocybin-active with 0.1–0.5 percent total tryptamines when dried.[114] More commonly reported as focus increases on the psilocybin-active species, this bitter-tasting Gymnopilus is unique for its strong anise fragrance, most noticeably emanating from the gills. It has a much thinner stem than the larger *Gymnopilus subspectabilis* and related species. Primordia that fail to mature and subsequently abort tend to show a bluing reaction more noticeably than healthier, vibrant primordia whose tissue is unbruised.

Gymnopilus purpuratus (COOKE & MASSEE) SINGER

CAP: 1–6 cm broad. Convex at first with an incurved then decurved margin, expanding with age to broadly convex to plane. Reddish purplish brown to purplish red with tinges of yellow and green, sometimes with bluish spots. Surface covered with scattered fibrillose/floccose patches. Context turning blue when cut. **GILLS:** Attachment adnexed to sinuate, waxy yellow becoming brownish cinnamon yellow with spore maturity. **STEM:** 20–40 mm long by 2–4 mm thick, stout, narrowing upward, with fine parallel longitudinal striations and covered with fine, fragile fibrils that are often dusted with rusty orange spores. Pallid when young, soon yellowish to reddish brown overall, with greenish and yellowish overtones, struck with bluish tones or bruising bluish. Flesh purplish. **MICROSCOPIC FEATURES:** Spores bright rusty orange in deposit, 7.5–9 by 4.8–5.7 µm, ellipsoid, slightly warty or with small ridges. Cheilocystidia cylindrical to ventricose, 18–26 by 4–4.5 µm, narrowing toward apex but often with a 3-µm capitate end. Narrowly saccate, 14–16 by 4–4.5 µm. Lacking pleurocystidia.

HABIT, HABITAT, AND DISTRIBUTION: Growing on rotting wood from downed trees to wood chips used for landscaping or bedding of animals. Reported from North and South America, United Kingdom, Europe, Australia, New Zealand, and Chile. Given these geographically distant locations, this species is probably circumpolar, resident in the diverse temperate forests of the world.

Gymnopilus purpuratus fruits from wood.

Cut stems of *Gymnopilus purpuratus*. Note strong bluing reactions.

COMMENTS: Weakly to moderately active. Jochen Gartz reported that dried, cultivated specimens yielded highly variable amounts from 0.07 to 0.33 percent psilocybin and psilocin and 0.005 to 0.05 percent baeocystin.[115] Later, Gartz found roughly similar amounts: 0.25 percent psilocybin, 0.33 percent psilocin, and 0.03 percent baeocystin—also from cultivated specimens.[116] The younger specimens contained more psilocybin and psilocin than the mature specimens.

Gymnopilus subspectabilis HESLER

Common Names: Big Gym, Giant Laughing Mushroom, Big Laughing Mushroom, Giant Gymnopilus

CAP: 5–40 cm broad. Convex to broadly convex, expanding to nearly plane with age. Bright yellowish orange becoming rusty orange to tawny gold to orangish brown or reddish brown at maturity. Surface dry, covered with fibrils or fibrillose scales. Margin incurved at first, when young can be decorated with remnants of the membranous partial veil, straightening or becoming wavy in age. Flesh yellowish. **GILLS:** Attachment adnate to sinuate to subdecurrent. Pale yellow to rusty orange, becoming rusty brown with spore maturity. Close to crowded. **STEM:** 30–250 mm long by 1–10 mm thick. Firm, solid, unequal, swelling in the middle or often narrowing toward the base. Rusty orange to yellowish orange, dingy brown toward the base. Surface covered with fine fibrils below the ring. Partial veil densely cortinate to membranous, usually leaving a well-formed, membranous annulus in the superior regions of the stem, soon dusted rusty orange from spores.

MICROSCOPIC FEATURES: Spores rusty orange in deposit, ovoid to ellipsoid, roughened, 7–10 by 4.4–6 µm. Pleurocystidia 21–37.3 µm long by 3.8–7.2 wide, fusoid-ventricose, narrowing toward the apex, scattered. Cheilocystidia fusoid-ventricose with subcapitate apices, 21–37 by 4–7 µm.

HABIT, HABITAT, AND DISTRIBUTION: Reported from southeastern and eastern North America. A fall mushroom, growing gregariously but most commonly in clusters on hardwood trees and stumps, often at the base of trees, from roots or occasionally from buried wood.

COMMENTS: Not recommended. Moderately psilocybin-active, *Gymnopilus subspectabilis* is extremely bitter. Because it has psilocybin and only intermittently psilocin, it will not necessarily bruise bluish when injured or cut—as it is the presence of psilocin that gives rise to bluing. Producing the largest fruitbodies of any psilocybin-containing mushroom, it has some close look-alikes. *Gymnopilus subspectabilis* is a synonym of what has been called *Gymnopilus spectabilis*, a species that in its many iterations has been reportedly psilocybin active and also neurotoxic,[117] although these original reports may have been based on misidentifications. Given the wide distribution of this species and the danger of misidentification, I would not depend on geographical zones for separating these species taxonomically by sight. *Gymnopilus junonius*, a related species is of concern, thought by some experts to contain psilocybin and the neurotoxic oligoisoprenoids gymnopilins, toxins that may also occur in other *Gymnopilus* species. It is best not to ingest them, especially since the smaller *Gymnopilus* species and *Galerina* species can appear similar to those not skilled in noting the differences between these two genera. *Psilocybe* species may be a much better choice for those wishing to ingest a psilocybin mushroom.

Gymnopilus subspectabilis, the Big Laughing Gym, is typically a large species that grows in clusters at the base of trees.

The Genus *Inocybe*

The genus *Inocybe* has more than 1000 species, the vast majority of which are not poisonous or toxic; several species are psilocybin active. *Inocybe* means "fiber cap," so named because of the stringy, tomentose fibers emanating outward from the center of a typically conic cap that is sometimes umbonate. The cap colors can range from white to dull brown to reddish brown to lilac.

Dozens of *Inocybe* species are poisonous, typically from muscarine, which causes sweating, salivation, and increased tearing (lacrimation) within 15 to 60 minutes of ingestion. At high enough doses and without medical treatment, the muscarine-producing Inocybes can be deadly, but fatalities are infrequent as these "fiber caps" are not sought out as good edibles. I am concerned that would-be psilocybin psychonauts will experiment with them. None are yet known to simultaneously contain muscarine and psilocybin, although I would not find this surprising.[118]

All *Inocybe* species are mycorrhizal, making this the only mycorrhizal genus to host psilocybin species. (All other genera hosting psilocybin species are saprophytic.) The spore color is typically a variety of shades of dull to deep brown. These mushrooms are typically small to medium in size, and none are sought out, to the best of my knowledge, as culinary edibles.

Stijve and Kuyper tested twenty *Inocybe* species with an emphasis on those specimens with greenish gray colors or that stain blue, especially at the base of the stem.[119] Six species tested positive for psilocybin: *Inocybe aeruginascens, Inocybe caerulata, Inocybe corydalina* var. *corydalina, Inocybe coelestium, Inocybe haemacta*, and *Inocybe tricolor*. In addition, a constellation of species, many not yet named, are centered around *Inosperma calamistratum* (formerly *Inocybe calamistrata*) can have a dark bluish black stem bases but has not tested positive for psilocybin or psilocin nor muscarine—yet (see page 71). We do not yet know what compounds are responsible for the bluing reaction in *Inosperma calamistrata.* I am listing only two representative psilocybin-active *Inocybe* species here.[120] I expect that dozens of psilocybin-active Inocybes are yet to be discovered and named. Please do not consume Inocybes because we do not have a good understanding of their complex biochemistries. Psilocybe and Panaeolus species are safer.

Inocybe aeruginascens BABOS

Common Name: Green Flushed Fiber Cap

CAP: 1.0–5.0 cm broad. Conic at first, expanding with age to convex and eventually plane with an obtuse umbo. Margin incurved when young, soon straightening. Surface adorned with radial fibrils, more floccose toward the disc. Color sordid buff to sordid ochraceous brown, often with greenish tinges. **GILLS:** Attachment adnate to nearly free, crowded, pale grayish brown to clay brown with greenish tones or bruising greenish where injured. **STEM:** 22–50 mm long by 3–7 mm thick.

Inocybe aeruginascens is a rare species currently known from the United Kingdom and Europe. Note light bluing on lower regions of the stems.

Equal to swelling at the base, solid, whitish to pallid at first, becoming bluish green from the base upward. Surface pruinose near the apex and longitudinally fibrillose below. Partial veil cortinate, soon disappearing. Flesh whitish, soon bruising bluish green, or naturally bluish green near the base. Odor disagreeable, soapy smelling. **MICROSCOPIC FEATURES:** Spores clay brown in deposit, smooth, ellipsoid, inequilateral, 7.5–10.0 by 4.0–5.0 µm. Pleurocystidia 31–71 by 9–24 µm, narrow to broadly fusiform, subclavate, with clear to yellowish tinged walls. Cheilocystidia scattered, when present similar to pleurocystidia.

HABIT, HABITAT, AND DISTRIBUTION: Reported from the British Isles and central Europe, an infrequently encountered species that is likely to be more widely distributed than currently reported. Found in sandy soils (including dunes), mixed conifer and deciduous forests, and underneath poplars (*Populus* spp.) and willows (*Salix* spp.) from June through October.

COMMENTS: Weakly to moderately active with a reported maximum of 0.28 percent psilocybin, 0.008 percent psilocin, and 0.08 percent baeocystin.[121] Also reported to contain 0.40 percent psilocybin, 0 percent psilocin, 0.52 percent baeocystin, and 0.35 percent aeruginascin.[122] Jochen Gartz claimed to have grown this species in vitro, apart from tree roots, which is surprising to me, as most mycorrhizal species are difficult to grow in culture. He reported the formation of bluish green sclerotia.

Inocybe corydalina QUÉLET = *Inocybe corydalina* var. *corydalina* QUÉLET

Common Name: Greenflush Fibercap

CAP: 3.8–5.2 cm broad. Obtusely conic to convex at first with an incurved and often denticulate margin, becoming broadly convex to nearly plane in age with or without a broad low umbo. Brown to buff brown, with greenish gray tones, and darker brown to greenish blue, sometimes nearly black, near the center. Surface covered with appressed fibrillose squamules, more densely toward the disc. Flesh white. **GILLS:** Attachment narrowly adnate, crowded, broad, with a minutely fringed margin. Pale brown to buff or pale grayish brown. **STEM:** 24–95 mm long by 5–15 mm thick. Solid, equal to enlarged near the base. Surface smooth above to fibrillose and longitudinally striate below. Whitish to dull gray below and grayish ochraceous brown to sordid brown overall, with base often with grayish greenish tinges. Flesh white to gray toward the base, slightly reddening upon exposure. Partial veil cortinate, soon disappearing. Scent aromatic, similar to Peruvian balsam. **MICROSCOPIC FEATURES:** Spores brown

Inocybe corydalina, the Greenflush Fibercap, contains a small amount of psilocybin and presumably psilocin, as it bruises lightly bluish or naturally in age.

in deposit, smooth, lemon to almond shaped, 7–10 by 5–6 µm. Pleurocystidia 33–70 by 9–21 µm, cylindrical to clavate-cylindrical. Cheilocystidia rare, when present similar to pleurocystidia.

HABIT, HABITAT, AND DISTRIBUTION: Widespread across Europe, the British Isles, and North America, August through October, primarily under deciduous trees, oak (*Quercus*), beech (*Fagus*), hornbeam (*Carpinus*), and, to a lesser degree, a conifer like the blue spruce (*Picea pungens*).

COMMENTS: Weakly active.[123] One variety, *Inocybe corydalina* var. *corydalina*, contains up to 0.032 percent psilocybin, 0 percent psilocin, and 0.034 percent baeocystin, whereas *Inocybe corydalina* var. *erinaceomorpha* has 0.10 percent psilocybin, 0 percent psilocin, and 0.034 percent baeocystin but typically does not feature bluish tones. *Inocybe corydalina* var. *erinaceomorpha* (Stangl & J. Veselský) Kuyper is now classified as its own species: *Inocybe erinaceomorpha* Stangl & J. Veselský.[124] This species has been reported from the eastern United States and central Europe. As Inocybe chemistry and taxonomy remains largely unexplored, this species is likely at a center of many species yet to be elucidated.

The Genus *Pluteus*

The genus *Pluteus* features members that are primarily wood decomposers with caps typically convex to plane, gills that are pink, free from the stem at maturity, and a ringless stem. Species of *Pluteus* have pinkish to salmon-colored to brownish spores. Many have pinkish tones in the cap as well. The mushrooms are generally fragile, easily broken when picked, and fall apart with minimal handling or impact. *Pluteus* is Latin for "movable shelter" or "roof" in reference to the overarching cap in some species.

Most are small to mid-sized and give pinkish to salmon to flesh-colored spore deposits. Only two of the psilocybin-active species are discussed here, *Pluteus phaeocyanopus* and *Pluteus salicinus*, but three additional *Pluteus* species are potentially psilocybin-active: *Pluteus glaucus*, *Pluteus septocystidatus*, and *Pluteus villosus*,[125] which was reported to have up to 0.28 percent psilocybin and 0.12 percent psilocin, a range comparable to that of *Pluteus salicinus*, which Kuyper reported at 0.25 percent.[126]

Pluteus phaeocyanopus mushrooms are exceptionally fragile and often break apart when harvesting. Note bluing on stems. *Pluteus phaeocyanopus* is psilocybin active, but weakly so.

Pluteus phaeocyanopus MINNIS & SUNDBERG

CAP: 0.5–2 cm broad. Conic convex, expanding to broadly convex plane, subumbonate. Surface dry, dull, covered with fine, velvety hairs, dark brown at the center, lighter toward the translucent-striate margin, incurved at first, rimose toward the edges at maturity. **GILLS:** Close, ascending, free, with three tiers of intermediate gills, pallid when young, becoming pale salmon brown at maturity. **STEM:** 10–30 mm long by 4 mm thick, narrowing toward the apex and swelling at the base, often curved, splitting longitudinally in maturity, with the surface fibrillose-striate, whitish overall but often bluish gray toward the base. **MICROSCOPIC FEATURES:** Spores salmon to pinkish brown, 6.2–8.8 by 5.3–7.5 µm, globose to broadly ellipsoid, ovoid in face view. Pleurocystidia 39–75 by 17–33 µm, dark broadly lageniform to utriform, with short necks and obtuse apices. Cheilocystidia 33–51 by 17–28 µm, pyriform to clavate to lageniform to utriform.

HABIT, HABITAT, AND DISTRIBUTION: Solitary to scattered, growing on humus and from buried roots in mixed forests of oak (*Quercus*), maple (*Acer*), ash (*Fraxinus*), and basswood (*Tilia*) along the west coast of North America from Southern California to British Columbia; also reported from Florida. Probably more widely distributed than currently reported.

COMMENTS: Although I know of no recent analysis of freshly picked specimens, this is a psilocybin-active species. The *Pluteus* species are typically fragile and the bluing stem bases can easily be missed if not carefully picked. *Pluteus phaeocyanopus* is reportedly thus far found only in the United States.

Pluteus salicinus (PERSOON ex FRIES) KUMMER

Common Names: Willow Shield, Knackers Crumpet

CAP: 3–7 cm broad. Cap convex to broadly convex, expanding with age to broadly convex to plane. Gray to gray greenish to bluish gray, darker toward the disc. Surface smooth to finely scaly near the center. **GILLS:** Free, not attached. Pallid to cream, soon pinkish to salmon colored at spore maturity. **STEM:** 40–100 mm long by 2–6 mm thick. White to grayish green, often with bluish tones. Flesh often bruising bluish where injured, especially near the base. Base of stem bruising bluish. **MICROSCOPIC FEATURES:** Spores pinkish in deposit, smooth, ellipsoid to ovoid, 7–8.5 by 5–6 µm. Pleurocystidia fusiform to lageniform, with or without hooked ends, 58–90 by 10–22 µm, with an apex 5–10 µm thick. Cheilocystidia pear shaped to clavate to cylindrical or slightly lageniform, 30–85 by 8–20 µm.

HABIT, HABITAT, AND DISTRIBUTION: Solitary to few, usually not in large colonies, growing on the humus below hardwood trees such as willow (*Salix*), alder (*Alnus*), beech (*Fagus*), oak (*Quercus*), eucalyptus (*Eucalyptus*), and poplar (*Populus*), or on their woody detritus; the species especially favors riparian habitats. Widely distributed across the United States, the British Isles (England, Scotland, and Wales), Europe, and Russia. Also collected from South Korea, India, New Guinea, and South Africa. Likely to be more broadly distributed than currently reported.

COMMENTS: I suspect that this species is at the center of a constellation of geographically dispersed psilocybin-active species in *Pluteus*. Analyses show *Pluteus salicinus* to be weakly to moderately active.[127,128] I do not know of any

Pluteus salicinus is a widely distributed species. Note bluing at the base of the stem. *Pluteus salicinus* is weakly psilocybin-active.

reports of people ingesting this fragile mushroom. Although pristine when found, this mushroom tends to fall apart when jostled in a collecting basket, if not from the mere act of plucking it from the ground; it is very fragile.

Stijve and Kuyper reported a range of 0.05–0.25 percent psilocybin, 0 percent psilocin, and 0–0.008 percent baeocystin.[129] A. L. Christiansen et al. found 0.35 percent psilocybin and 0.011 percent psilocin.[130] See also Saupe (1981) and Stijve and Bonnard (1986).

New Species? To Be Determined . . .

Many new psilocybin-active species are being discovered or, in the case of prior Indigenous use, rediscovered by academic mycologists. Some of these mushroom species may have new Latin bi-nomials subsequent to the submission of this book. I am highlighting a few that I find particularly exciting here. To stay updated, please go to www.psilocybinmushroomsintheirnaturalhabitats.com.

Psilocybe gandalfiana nom. prov. is a secotioid psilocybin-active species growing near Lake Tahoe, California.

A *Psilocybe* species often called *Psilocybe tasmaniana* but may be a different species. Found in New Zealand.

Likely a new *Psilocybe* species, "Psilocybe sp-CA03," growing in wood chips used for landscaping in Pasadena, California.

Psilocybe polytrichoides, a new, secotioid species close to *Psilocybe pelliculosa*, growing in the Mount Shasta region of Northern California.

Psilocybe incaica nom.prov. a potentially new *Psilocybe* species from Peru, lower right.

Acknowledgments

María Sabina, Tina Wasson, and Catherine "Kit" Scates were three great mycologists whose stewardship as knowledge keepers greatly influenced my life as a mycologist. This book is dedicated to them.

Many others have helped me on this long journey, the foremost of which has been Michael Beug, Andrew Weil, Alexander Smith, Daniel Stuntz, and Gastón Guzmán.

Others who have contributed on the path to make this book possible are Ahmed Abdel-Azeem, Donald Abrams, John W. Allen, Julio Alvarez, Mike and Eileen Amaranthus, Stephen Apkon, Geeta Arora, Linsie Auseth, Don José Ávila, Anthony Back, Erlon Bailey, Zolton Bair, Chase Beathard, Denis Benjamin, Ann Beug, Tom Bigelow, Jeremy Bigwood, Robert Blanchette, Eugenia Bone, Jan Borovička, Anthony Bossis, Betsy Bullman, Ignacio Seral Bozal, Alexander Bradshaw, Taylor Bright, David Bronner, Bernard Brown, Peter Buchanan, Tad and Emilia Buchanan, Betsy Bullman, Seymour Burgess, Jon Callaghan, Kyle Canan, Robin Carhart-Harris, Steve Cividanes, James Conway, J.A. Cooper, Alexandra Cohen, Katsie Cook, Ashley Cowie, John Cumbers, Lavinia Currier, Alfred DH, Renee Davis, Wade Davis, Chelsea DeLeon, Bryn Dentinger, Dennis Desjardin, Felice Di Palma, Rick Doblin, Gül Dölen, Sanjay Dubé, Josh Dugdale, James Fadiman, Amanda Feilding, Rock Feilding, Cosmo Feilding-Mellen, Tim Ferris, Giuliana Furci, Richard Gaines, Adam Gazzaley, Inti Garcia Flores, Tim Girvin, James Gouin, Alex and Allyson Grey, Roland Griffiths, Charles Grob, Hans Grootewal, Guujaaw, Gaston Guzmán, Laura Guzmán-Dávalos, Marcina Hale, Graham and Santha Hancock, Michael Hanuschik, Kalin Harvey, David Hawksworth, Mark Henson, David Hibbett, Patrick C. Hickey, Paxton Hoag, Dave Hodges, Julie Holland, William Hyde, Baba Kilindi Iyi, Jim Jacobs, James & Zem Joaquin, Chris Kantrowitz, Don Kneeland, Daniel Kraft, Paul Kroeger, Kim Kuypers, Nikola Lačković, Rick Langer, Bruce Langereis, Renée Lebeuf, Andrew Lenzer, Bill Linton, Rich Locus, Mike Magee, Geraldine Manson, Albert and Murdena Marshall, Joel McCleary, Dennis McKenna, Terence McKenna, Andy MacKinnon, Jonathan Meader, Poncho Meisenheimer, Kyle Meyer, Nick Moore, Steve Morris, T. Moult, Casey Mullen, Iván Pérez Muñoz, Brian Muraresku, Patrick Murphy, Regan Nally, Machiel E. Noordeloos, Musa Ngwenya, Chris W. Nelson, Melissa Nelson, Paul Noth, David Nutt, Michael Olson, Scott Ostuni, Jonathan Ott, Chris Park, Michael Pollan, Scott Redhead, Benhadja Ahmed Riadh, Bill Richards, Marlena Robbins, Joey Rootman, Martine and Bina Rothblatt, Nathan Sackett, Giorgio Samorini, Eesmyal Santos-Brault, Ethan Schaffer, Steven Schnoor, Nicolas Schwab, Christian Schwarz, Cosmo Sheldrake, Merlin Sheldrake, Alexander Sherwood, Sasha and Ann Shulgin, Josh Siegel, João Silva, Mike Sinyard, Autumn Skye, Ryan Snyder, Bulmaro Solano, Azureus Stamets, LaDena Stamets, Lilly Stamets, Patricia Stamets, William K. Stamets, Lee and June Stein, David Sumerlin, Gerhard Suttner, David Tatelman, David Taylor, Santiago Tiscornia, D. B. Townsend, Jim Trappe, Tina Trujillo, Nancy Turner, Sara and Steve Urquhart, Rytas Vilglays, Michael Wallace, Zach Walsh, Roy Watling, Bill Webb, Will Weisman, Annie Weissman, Clint Werner, Dakota Wint, Dusty Yao and deep gratitude to the prolific Alan

Rockefeller, also an excellent photographer, and the many other skilled photographers and artists who provided images. Zolton Bair, Jim Fadiman, Pamela Kryskow, Mark Plotkin, Tom Rieldlinger, and Bill Stamets are thanked for their time reading, editing, and for proofreading.

The mix of musicians who have accompanied me on this journey include Adam Bauer, Beats Antiques, Buddha Bar contributors, Burning Spear, Crosby, Stills & Nash, Amani & Desert Dwellers, Bob Dylan, Pink Floyd, Marvin Gaye, Grateful Dead and The Dead, JL & Afterman, Kaya Project, Majik Band, Bob Marley, Dave Matthews, Paul McCartney and The Beatles, Joni Mitchell, Van Morrison, Bob Moses, Kacey Musgraves, Nasiri, Willie, Micah and Lukas Nelson, 1 Giant Leap, Ott, Peter Tosh, Phutureprimitive, Radiohead, The Rolling Stones, Sting, James Taylor, The Human Experience, U2, Neil Young, YAIMA . . . and so many more!

Thank you, Julie Bennett, Ashley Pierce, Lizzie Allen, Philip Leung, Joey Lozada, and Chloe Aryeh of Ten Speed Press; copyeditor Lisa Brousseau, who masterfully edited this book; proofreader Sasha Tropp; and David Arora, who advocated my work to Phil Wood, the founder of Ten Speed Press. Perhaps most importantly, I am deeply appreciative of my older brother John Stamets, who initially steered my lifelong journey toward exploring psilocybin mushrooms. I will forever be grateful to him and, by extension, to all of you. Together, we continue the lineage of being psilocybin mushroom wisdom keepers.

Glossary

acute pointed, sharp

adnate (of gills) bluntly attached to the stem

adnexed (of gills) attached to the stem in an ascending manner

Agaricaceae a family of fungi primarily comprised of members of the genus *Agaricus*

Agaricales the order that includes all mushrooms with true gills

agarics mushrooms with gills

ampullaceous shaped like a bottle or bladder

amyloid an aggregate of proteins in the flesh or spores of a mushroom that exhibits a characteristic bluish reaction in Melzer's iodine

annular (on the stem) resembling a ring

annulate having an annulus

annulus tissue remnants of the partial veil adhering to the stem and forming a membranous ring

apex (of the stem) the top or highest point or region

appendiculate (of the cap margin) hanging with veil remnants

appressed flattened

ascending (of the gills) gills extend upward from the margin of the cap to their attachment at the stem

Ascomycetes fungi that produce spores by an ascus, a sac-like cell (as opposed to the basidium of Basidiomycetes)

aspect general shape or outline of a mushroom

attachment connection between the gills and stem

atypical not typical

autonomous being of its own; an entity in itself

basidiocarp the fruitbody, the fleshy reproductive structure, that produces basidia

Basidiomycetes fungi that bear spores upon a basidium (as opposed to the Ascomycetes, which bear spores in an ascus)

basidium (pl. basidia) a particular fertile cell in which meiosis occurs and by which spores are produced

broad a relative term connoting width (as opposed to length) in reference to the gills (in the order narrow, moderately broad, broad)

buff dingy yellowish brown

campanulate (of the cap) bell shaped

cartilaginous brittle, not pliant

caulocystidia sterile cells on the stem

cellular composed of globose to generally rounded cells, not thread-like

cespitose growing clustered from a common base

cheilocystidia sterile cells on the gill edge; sometimes called marginal cystidia

clamp connection an elbow-like protuberance that arches over the walls between cells in dikaryotic mycelium of some mushroom species

clavate blunted in shape

claviform club shaped

close (of the gills) a relative term in reference to the spacing of the gills (in the order crowded, close, subdistant, distant)

collyboid resembling a mushroom from the genus *Collybia*, typically with expanded caps (broadly convex to plane) and stems not more than two or three times the breadth of the caps, often growing in groups or clusters; term is often used in contrast to mycenoid

colonization (biological definition) occupation of a habitat or territory by a biological community or of an ecological niche by a single population of a species, including species of microbes (bacteria, archaea, and fungi) to more complex organisms, like plants and animals

colonization (cultural definition) the extension of colonial state power through cultural knowledge, activities, and institutions (particularly education and media) or the systematic subordination of one conceptual framework or cultural identity over others

complex a cluster of similarly related species typified by a central species

concolorous having the same color

conic (of the cap) shaped like a cone

conspecific the same as, synonymous with

context the flesh of the cap or stem

convex (of the cap) rounded, like the outside half of a sphere

Copelandian of or relating to the genus *Copelandia*, distinguished from its close relatives, the *Panaeolus* species, by their dung-loving habitat, strong bluing reaction, and large pleurocystidia

coprophilous growing on dung

cortinate a type of partial veil consisting of fine cobweb-like threads

crowded (of the gills) a relative term used in describing the narrow spacing of the gills (in the order crowded, close, subdistant, distant)

cuspid (adj. cuspidate) sharp, pointed

cuticle the surface layer of cells on the cap that can undergo varying degrees of differentiation

cystidia microscopic sterile cells

deciduous trees and other plants that seasonally shed their leaves

decurrent (of the gills) type of attachment where the gills markedly run down the stem

decurved (of the margin) the shape or curvature bends directly downward

denticulate tooth-like, lined with triangular fragments of tissue

dextrinoid turning reddish brown upon application of Melzer's reagent

dichotomous repeatedly splitting or forking in pairs

dikaryophase phase in which there are two individual nuclei in each cell of the mushroom

dikaryotic state of cells in the dikaryophase

diploid genetic condition in which each cell has the full set of chromosomes for the species (2N)

disc central portion of the cap

distant relative term implying the broad spacing of the gills (in the order crowded, close, subdistant, distant)

elevated (of the margin) describing the type of cap whose margin is uplifted, usually seen in age

ellipsoid shaped like an oblong circle

equal (of the stem) evenly thick

eroded (of the cap margin or gills) irregularly broken

evanescent fragile and soon disappearing

felted covered in densely packed fibers

fibril fine, delicate strand found on the surface of the cap or stem

fibrillose having fibrils

fibrous (of the stem) composed of tough, string-like tissue

filamentous composed of hyphae or thread-like cells, which may undergo gelatinization

flexuose, flexuous (of the stem) bent alternately in opposite directions

floccose, flocculose easily removed; usually referring to wooly tufts or cottony veil remnants on the cap or stem

fugacious impermanent, easily torn or destroyed

fusiform spindle shaped, tapering at both ends

fusoid (of the cystidia) rounded and tapering from the center

fusoid-ventricose swollen in the middle and narrowing toward the ends

gelatinous having the consistency of a jelly, usually translucent

glabrescent becoming glabrous

glabrous smooth, bald

glutinous having a highly viscous gelatinous layer, an extreme condition of viscidity

gregarious growing in numerous to dense groups but not clustered as in cespitose

group all the related varieties of one species

habit the way in which mushrooms are found growing (whether singly or numerous to clustered from a common base) and the aspect of their forms

habitat substrate in which the mushrooms are found

hemispheric (of the cap) resembling a hemisphere

holotype a single specimen that is designated in the original description

hyaline appearing glassy or transparent

hygrophanous markedly fading in color when drying

hypha (pl. hyphae) individual cells of the mycelium

hypogeous growing below the surface of the ground

incurved (of the margin) curved inward

indigenous native to a certain region

inequilateral unsymmetrical or lopsided

KOH chemical symbol for potassium hydroxide, an agent commonly used to revive dried mushroom material for microscopic study at a concentration of 2.5 percent (2.5 g of KOH in 97.5 mL of distilled water)

lageniform broad below and tapering to a slender neck above

lamellae gills, the spore-producing, plate-like structures radially emanating underneath and from the center of the cap

lanceolate tapering to a point at the apex and sometimes at the base, like a leaf

lecythiform shaped like a bowling pin or a bottle

lignicolous growing in wood or on a substratum composed of decayed wood

lignin basic organic substance of woody tissue other than cellulose

lubricous slightly slippery but not viscid

macroscopic visible to the naked eye

meiosis process of reduction division by which a single cell with one nucleus divides into four cells with one nucleus apiece; each nucleus has one-half the genetic material of the parent cell

membranous thin, tissue-like, as in a homogeneous membrane

micron (μm) one-millionth of a meter, one-thousandth of a millimeter

microscopic visible only with the aid of a microscope

mottled spotted, as from the uneven ripening of the spores in the genus *Panaeolus*

mucronate tipped with an abrupt short point

mycenoid resembling a mushroom from the genus *Mycena*: tall, slender mushrooms with long stems and comparatively small, conic caps; term is often used in contrast to collyboid

mycology the study of fungi

myconaut person who ingests mushrooms

mycorrhizal unique type of symbiotic relationship a fungal mycelium may form with the roots of a seed or spore-producing plants

nomenclature any system of classification

nucleus (pl. nuclei) concentrated mass of differentiated protoplasm in all cells that plays an integral role in the reproduction and continuation of genetic information to daughter cells

obtuse blunt (as opposed to pointed)

ochraceous light orangish brown to pale yellow brown

olivaceous olive gray brown

ovoid oval to egg shaped

pallid very pale in color, almost a dull whitish

papilla nipple

papillate having a nipple

partial veil inner veil of tissue extending from the cap margin to the stem and at first covering the gills

pellicle a skin-like covering on the cap, often gelatinous and separable

persistent not deteriorating with age; present throughout the life of the fruitbody

phototropic attracted or sensitive to light

pileocystidia a sterile cell or body, frequently with a distinctive shape, occurring on the surface of the mushroom pileus

pileus cap of the mushroom

pith central stuffing of stems of some mushrooms

pleurocystidia sterile cells on the surface of the gills; sometimes called facial cystidia

pliant flexible

pore circular depression evident on the end of spores in many species

pruinose having the appearance of being powdered due to an abundance of caulocystidia on the stem surface

pseudorhiza long root-like extension of the lower stem

psilocin 4-hydroxy-N,N-dimethyltryptamine, the active yet unstable compound that docks with neuroreceptors, which, in quantity, alters consciousness

psilocybin 3-[2-(dimethylamino) ethyl]-1H-indol-4-yl dihydrogen phosphate, also known as 4-phosphoryloxy-N,N-dimethyltryptamine, the active, relatively stable, prodrug to psilocin and the principal psychoactive molecule in many species of mushrooms

psychonaut person who ingests psilocybin mushrooms for spiritual voyages

pyriform pear-shaped

reticulate marked by lines, usually parallel

rhizomorphs cord-like strands of twisted hyphae present around the base of the stem

rhomboid resembling a rhombus

rimose having many clefts, cracks, or fissures

rostrate having a beak or beaklike process

seceding (of the gills) condition where the gills have separated in their attachment to the stem and have the appearance of being free; often leaving longitudinal lines on the stem where the gills once were connected

secotioid intermediate growth stage where the cap of the fruitbody is bag-like, covering the developing spore-producing surfaces (usually gills) in the evolution toward becoming a gasteroid form

sinuate (of the gills) type of attachment that seems to be notched before reaching the stem

sordid dingy looking

spores reproductive cells or "seeds" of fungi borne on specialized cells

squamule a small scale

squamulose covered with small scales

Stametsian of and related to the mycology philosophy and methods of Paul Stamets

stipe stem of a mushroom

striate having stripes, lines, or grooves,

strigose having long stiff hairs

Strophariaceae family of mushroom containing the closely allied genera that historically contained *Hypholoma, Pholiota, Psilocybe*, *Stropharia*, and nearly a dozen other genera

subclose in reference to the spacing of the gills, between close and crowded

subdistant in reference to the spacing of the gills, between close and distant

subequal (of the stem) not quite equal

substratum the substance in which mushrooms grow

superior term used to designate the location of the annulus in the upper third of the stem

tawny approximately the color of a lion

taxa a single taxonomic category, such as a species or variety

terrestrial growing on the ground

translucent transmitting light diffusely, semi-transparent

translucent-striate typically used to describe the visibility of the underlying gill plates when viewing a moist mushroom cap from above

umbilicate (of the cap) depressed in the center

umbo knob-like protrusion in the center of the cap

umbonate having an umbo

uncinate type of gill attachment with a hooked appearance

undulating wavy; the cap margin of *Psilocybe cyanescens* is a classic example

universal veil an outer layer of tissue enveloping the cap and stem of some mushrooms, best seen in the youngest stages of development

utriform shaped like a womb

variety a subspecific epithet used to describe a consistently appearing variation of a particular mushroom species

veil a tissue covering mushrooms as they develop; also see definitions of partial veil and universal veil

ventricose swollen or enlarged in the middle

vesiculose covered in small, fluid-filled sacs

vinaceous red wine in color

violaceous bluish purple in color

viscid slimy, slippery, or sticky to the touch; enhanced in moist conditions

Notes

Introduction: The Psilocybin Mushroom Journey Begins

1. Smith, A. H. 1973. *Mushrooms in Their Natural Habitats.* Hafner Press, New York.
2. Ainsworth, G. C. 1976. *Introduction to the History of Mycology.* Cambridge University Press, New York.
3. Margolin, M., Hartman, S. 2021. Jews, Christians, and Muslims are reclaiming ancient psychedelic practices, and that could help with legalization. *Rolling Stone*, April 23.
4. Hawksworth, D. L. Lücking, R. Fungal. 2017. Diversity revisited: 2.2 to 3.8 million species. *Microbiol Spectrum* 5(4). doi:10.1128/microbiolspec.funk-0052-2016.
5. BC Centre for Disease Control. Fact sheet, wild mushrooms may be poisonous. http://www.bccdc.ca/resource-gallery/Documents/Educational%20Materials/EH/FPS/Fruit%20and%20Veg/Wild%20Mushrooms%20May%20Be%20Poisonous%20Fact%20Sheet.pdf.
6. Colorado State University. Fact sheet, food source information: mushrooms. https://www.chhs.colostate.edu/fsi/food-articles/produce-2/mushrooms.
7. Bradshaw, A. J., Ramírez-Cruz, V., Awan, A. R., et al. 2024. Phylogenomics of the psychoactive mushroom genus *Psilocybe* and evolution of the psilocybin biosynthetic gene cluster. *Proceedings of the National Academy of Sciences USA* 121(3):1–9. doi.org/10.1073/pnas.2311245121.
8. Guzmán, G., Allen, J.W., Gartz, J. 1998. A worldwide geographical distribution of the neurotropic fungi, an analysis and discussion. *Annali del Museo Civico di Rovereto* 14:189–280.
9. Strauss, D., Ghosh, S., Murray, Z., et al. 2022. An overview on the taxonomy, phylogenetics and ecology of the psychedelic genera *Psilocybe, Panaeolus, Pluteus,* and *Gymnopilus. Frontiers in Forests and Global Change* 5:813998. doi:10.3389/ffgc.2022.813998.
10. Strauss et al. 2022.
11. Stijve, T., Kuyper, T. W. 1985. Occurrence of psilocybin in various higher fungi from several European countries. *Planta Medica* 51(5):385–387. doi.org/10.1055/s-2007-969526.
12. Gotvaldová, K., Borovička, J., Hájková, K., et al. 2022. Extensive collection of psychotropic mushrooms with determination of their tryptamine alkaloids. *International Journal of Molecular Sciences* 23(22):14068. doi:10.3390/ijms232214068.
13. Besl, H. 1993. *Galerina steglichii* spec. nov., ein halluzinogener Häubling. *Zeitschrift für Mykologie* 59(2):215–218.
14. Tanaka, M., Hashimoto, K., Okuno, T., Shirahama, H. 1993. Neurotoxic oligoisoprenoids of the hallucinogenic mushroom, *Gymnopilus spectabilis. Phytochemistry* 34(3):661–664. doi.org/10.1016/0031-9422(93)85335-O.
15. Lenz, C., Dörner, S., Sherwood, A., Hoffmeister, D. 2021. Structure elucidation and spectroscopic analysis of chromophores produced by oxidative psilocin dimerization. *Chemistry* 27(47):12166–12171. doi.org/10.1002/chem.202101382.
16. Patocka, J., Wu, R., Nepovimova, E., et al. 2021. Chemistry and toxicology of major bioactive substances in *Inocybe* mushrooms. *International Journal of Molecular Sciences* 22(4):2218. doi: 10.3390/ijms22042218.
17. The median lethal dose (LD50) for psilocybin is 280 mg/kg, or 19.6 g psilocybin for a 70 kg person. At 1 percent psilocybin in a dried mushroom, this would be approximately 1.96 kg of dried mushrooms or 19.6 kg of fresh mushroom at 90 percent, 10 percent dried mass. This is the equivalent of a 154-pound person consuming 41.8 pounds of fresh psilocybin mushrooms—an impossible task. See https://www.aatbio.com/resources/toxicity-lethality-median-dose-td50-ld50/psilocybin.

Chapter 1. Evolution and Historical Use of Psilocybin Mushrooms

1. Bradshaw et al. 2024.
2. Meyer, M., Slot, J. 2023. The evolution and ecology of psilocybin in nature. *Fungal Genetics and Biology* 167:103812. doi.org/10.1016/j.fgb.2023.103812.
3. Bradshaw et al. 2024.
4. Smith, D. 1996. *The Cattle Egret* (Bubulcus ibis)*: Colonizer of Old-World Origin and a Vector of* Psilocybe cubensis *Spores.* Stain Blue Press, Spring, Texas.
5. Meyer, M. and Slot, J. 2023. The evolution and ecology of psilocybin in nature. *SSRN* Electronic Journal 10.2139/ssrn.4384673.
6. Reynolds, H. T., Vijayakumar, V., Gluck-Thaler, E., et al. 2018. Horizontal gene cluster transfer increased hallucinogenic mushroom diversity. *Evolution Letters* 2:88–101. doi.org/10.1002/evl3.42.
7. Pollan, M. 2002. *The Botany of Desire: A Plant's-Eye View of the World.* Random House, New York.
8. Hilden, N. 2021. Future space travel might require mushrooms. *Scientific American Space & Physics* 4(5). www.scientificamerican.com/article/space-travels-most-surprising-future-ingredient-mushrooms

9. Hanson, A. M., Hodge, K. T., Porter, L. M. 2003. Mycophagy among primates. *Mycologist* 17(1):6–10. doi.org/10.1017/S0269-915X(03)00106-X.

10. de Vos, C. M. H., Mason, N. L., Kuypers, K. P. C. 2021. Psychedelics and neuroplasticity: a systematic review unraveling the biological underpinnings of psychedelics psychiatry. *Frontiers in Psychiatry* 12:724606. doi.org/10.3389/fpsyt.2021.724606.

11. van der Merwe, B., van der, A., Rockefeller, A., et al. 2024. A description of two novel *Psilocybe* species from southern Africa and some notes on African traditional hallucinogenic mushroom use. *Mycologia*, 116(5), 821–834. doi.org/10.1080/00275514.2024.2363137.

12. Nair, J. J., Van Staden, J. 2014. Traditional usage, phytochemistry and pharmacology of the South African medicinal plant *Boophone disticha* (L.f.) Herb. (Amaryllidaceae). *Journal of Ethnopharmacology* 151(1).

13. Samorini, G. 1992. The oldest representations of hallucinogenic mushrooms in the world (Sahara Desert, 9000-7000 B.P.). *Integration* 2(3):69–78.

14. Samorini, G. 1995. Uso tradizionale di funghi psicoattivi in Costa d'Avorio? *Eleusis* 1:22–27.

15. Lajoux, J. D. 1963. *The Rock Paintings of Tassili.* The World Publishing Company, Cleveland, Ohio. Translated from Societe Nouvelle des Editions due Chene Paris, 1962: 68–73.

16. Wasson, R. G., Hofmann, A., Ruck, C. 2008. *The Road to Eleusis: Unveiling the Secret of the Mysteries.* North Atlantic Books, Berkeley, California.

17. Muraresku, B. C. 2020. *The Immortality Key: The Secret History of the Religion with No Name.* St. Martin's Press, New York.

18. Akers, B. P., et al. 2011. A prehistoric mural in Spain depicting neurotropic *Psilocybe* mushrooms? *Economic Botany* 65:121–128.

19. Wasson, R. G., Hofmann, A., Ruck, C. 2008. *The Road to Eleusis: Unveiling the Secret of the Mysteries.* North Atlantic Books, Berkeley, California.

20. Ott, J. 1993. *Pharmacotheon.* Natural Products Co., Kennewick, Washington.

21. Riedlinger, T. J. 2002. Polydamna's drug: Egyptian beer and the kykeon of Eleusis. *The Entheogen Review* 11(2):49–57.

22. Wasson, R. G. 1971. *Soma-divine mushroom of immortality.* Harcourt Brace Jovanovich, New York.

23. Maillart-Garg, M., Winkelman, M. 2019. The "Kamasutra" temples of India: A case for the encoding of psychedelically induced spirituality. *Journal of Psychedelic Studies* 3(2):81–103. doi.org/10.1556/2054.2019.012.

24. Battin, J. 2010. Le feu Saint-Antoine ou ergotisme gangreneux et son iconographie médiévale [Saint Anthony's Fire or gangrenous ergotism and its medieval iconography]. *Histoire des sciences médicales* 44(4):373–382.

25. Omissi, A. 2016. The cap of liberty: Roman slavery, cultural memory, and magic mushrooms. *Folklore* 127(3):270–285. doi.org/lo.1080/0015587X.2016.1155371.

26. Wint, Dakota. 2024. Personal communication with the author.

27. Wasson, R. G., O'Flaherty, W. D. 1982. The last meal of the Buddha. *Journal of the American Oriental Society* 102(4):591–603. doi.org/10.2307/601968.

28. Brown, J. B., Brown, J. M. 2016. *The Psychedelic Gospels: The Secret History of Hallucinogens in Christianity.* Park Street Press, Rochester, New York.

29. Curiously, I can trace my mother's lineage to the north clan of the Guilford House of Azure, a few miles from Stonehenge. I did not know this when I named my son Azureus and coauthored the naming of *Psilocybe azurescens.*

30. Sayin, H. U. 2014. The consumption of psychoactive plants during religious rituals: the roots of common symbols and figures in religions and myths. *NeuroQuantology* 12(2):276–296. doi.org/10.14704/nq.2014.12.2.753.

31. Shukir Muhammed Amin, O. 2017. Godess Kubaba. World History Encyclopedia. https://www.worldhistory.org/image/7198/goddess-kubaba/#google_vignette.

32. Berlant, S. R. 2005. The entheomycological origin of Egyptian crowns and the esoteric underpinnings of Egyptian religion. *Journal of Ethnopharmacology* 102:275–288. doi.org/10.1016/j.jep.2005.07.028.

33. El Enshasy, H., Elsayed, E. A., Aziz, R., et al. 2013. Mushrooms and truffles: historical biofactories for complementary medicine in Africa and in the Middle East. *Evidence-Based Complementary and Alternative Medicine* 2013:620451. doi.org/10.1155/2013/620451.

34. Budge, E. A. W. 1967. *The Egyptian Book of the Dead: The Papyrus of Ani.* Dover, Mineola, New York.

35. Abdel-Azeem, A. 2017. Egypt's national fungus day. *Microbial Biosystems* 2:21–25.

36. Berlant, S. R. 2005. The entheomycological origin of Egyptian crowns and the esoteric underpinnings of Egyptian religion. *Journal of Ethnopharmacology* 102:275–288. doi.org/10.1016/j.jep.2005.07.028.

37. Froese, T., Guzmán, G., Guzmán-Dávalos, L. 2016. On the origin of the genus *Psilocybe* and its potential ritual use in ancient Africa and Europe. *Economic Botany* 70(2):103–114.

38. Abdel-Azeem, A. M., Blanchette, R. A., Mohesien, M. T., et al. 2016. The conservation of mushroom in ancient Egypt through the present. *Proceedings of the First International Conference on Fungal Conservation in the Middle East and North of Africa*, 18–20 October 2016:1–2. Ismailia.

39. Abdel-Azeem, A. M. 2010. The history, fungal biodiversity, conservation, and future perspectives for mycology in Egypt. *IMA Fungus* 1(2):123–142.

40. Kilindi, I. 2013. "Breaking Convention Conference. High-Dose Mushrooms Beyond the Threshold." https://www.youtube.com/watch?v=ejdKeghBhNs: 27:09.

41. Finkelstein, I., Langgut, D. 2014. Dry climate in the Middle Bronze I and its impact on settlement patterns in the Levant and beyond: new pollen evidence. *Journal of Near Eastern Studies* 73(2):219–234. doi.org/10.1086/677300.

42. Rätsch, C. 1996. *Urbock, Beer Beyond Hops and Malts: From the Magic Potions of the Gods to the Psychedelic Beers of the Future.* AT Verlag, Aarau, Switzerland.

43. Siméon, R. 1885. *Dictionnaire de la langue nahuatl ou mexicaine.* Imprimerie Nationale, Paris.

44. Dibble, C. E., Anderson, A. J. O. 1950–1982. Translation and introduction of Bernardino de Sahagùn, *Historia General de Las Cosas de La Nueva España [Florentine Codex: General History of the Things of New Spain]*. 12 vols. University of Utah Press, Salt Lake City.

45. "They transformed agricultural fields into pastures by eating the vegetation, creating favorable conditions for Old World grasses and plants accidentally introduced by ships to replace native species. Cattle modified environments in ways that inadvertently benefited Spanish colonization." Ficek, R. E. 2019. Cattle, capital, colonization: tracking creatures of the Anthropocene in and out of human projects. *Current Anthropology* 60:S260–S271. www.journals.uchicago.edu/doi/pdf/10.1086/702788.

46. McIlvaine, C. 1900. *One Thousand American Fungi.* The Bowen-Merrill Co., Indianapolis, Indiana.

47. Schultes, R. E., Reko, P. 1938. *Collections of Economic Botany* No. 5548. Botanical Museum, Harvard University, Cambridge.

48. Johnson, J. B. 1939. The elements of Mazatec witchcraft. *Etnologiska Studier* 9:128–150.

49. Hofmann, A. 1980. The sacred mushroom teonanácatl: the Mexican relatives of LSD. In *LSD, My Problem Child.* McGraw-Hill, New York.

50. Pike, E., Cowan, F. 1959. Mushroom ritual versus Christianity. *Practical Anthropology* 6(5):145–151.

51. Schultes, R. E. 1940. Teonanacatl: the narcotic mushroom of the Aztecs. *American Anthropologist* (n.s.) 42(3):429–443.

52. Wasson, R. G. 1963. *The Hallucinogenic Mushrooms of Mexico and Psilocybin: A Bibliography.* 2nd ed. with corrections and addenda. Botanical Museum Leaflets, Harvard University 20(2a):25–73c. http://www.jstor.org/stable/41762224.

53. Allen, J. W. 1997. Wasson's first voyage: the rediscovery of entheogenic mushrooms. In *Mushroom Pioneers.* Psilly Publications and Raver Books, Seattle: 29. Also, see https://erowid.org/plants/mushrooms/mushrooms_article5.shtml

54. Wasson, R. G., Wasson, V. P. 1957. The riddle of the toad and other secrets mushroomic. *Mushrooms, Russia and History.* Pantheon Books, New York. Available via www.newalexandria.org/archive/mushrooms+russia+and+history+Volume+1.pdf

55. Riedlinger, T. J. 2024. Personal communication with the author.

56. Wasson, R. G. 1957. Seeking the Magic Mushroom. *Life Magazine*, May 13: 102, 109.

57. Wasson, R. G. 1957: 100–120.

58. "J.P. Morgan & Co. (see Wasson file)." MKUltra Subproject, no. 58 (doc: 17457). Washington, D.C.: National Security Archive at George Washington University. Uncovered by John Marks, a journalist specializing in CIA culture and history. Also, see Marks, J. 1979. *The Manchurian Candidate: The CIA and Mind Control.* Times Books, New York.

59. Riedlinger, T. J., ed. 1990. *The Sacred Mushroom Seeker: Tributes to R. Gordon Wasson by Terence McKenna, Joan Halifax, Peter T. Furst, Albert Hofmann, Richard Evans Schultes, and Others*: 215. Dioscorides Press, Portland, Oregon.

60. Wasson, V. P. 1957. I ate the sacred mushrooms. *This Week Magazine* (*Salt Lake Tribune*), May 19: 8.

61. Singer, R. 1949. The agaricales (mushrooms) in modern taxonomy. *Lilloa* 22: 474, 506. Chronica Botanica, Leyden. Note that Singer did not state "teonanacatl" but stated "Practical Importance: At least one species is used as a drug in Mexico (causing a temporary narcotic state of hilarity) but is poisonous when used in excess." Of course, he was wrong about Psilocybes being poisonous. They are not.

62. Singer, R., Smith, A. H. 1958. Mycological investigation on teonanácatl, the Mexican hallucinogenic mushroom. Part II. A taxonomic monograph of *Psilocybe*, section Caerulescentes. *Mycologia* 50:262–303.

63. Ott, J. 1993. *Pharmacotheon: Entheogenic Drugs, Their Plant Sources and History.* Natural Products Company: 302.
64. Heim, R. 1958. Diagnose latine du *Psilocybe wassonii* Heim, espece hallucinogene des Azteques. *Revue de Mycologie* 23(1):119–120.
65. Singer, R., Smith, A. H. 1958. New species of *Psilocybe. Mycologia* 50:141–142.
66. Singer, R. 1958. Mycological investigations on teonanácatl, the Mexican hallucinogenic mushroom. Part I. The history of teonanácatl, field work and culture work. *Mycologia* 50:239–261.
67. Singer, R. Smith, A. H. 1958. Mycological investigation on teonanácatl, the Mexican hallucinogenic mushroom. Part II. A taxonomic monograph of *Psilocybe*, section Caerulescentes. *Mycologia* 50:262–303.
68. Heim, R., Cailleux, R. 1960. Nouvelle contribution a la Connaissance des *Psilocybes* Hallucinogenes du Mexique. *Revue de Mycologie* 24:437–441.
69. Bartlett, C., Marshall, M., Marshall, A. 2012. Two-eyed seeing and other lessons learned within a co-learning journey of bringing together Indigenous and mainstream knowledges and ways of knowing. *Journal of Environmental Studies and Sciences* 2:331–340.
70. Paul Stamets received the MSA Wasson Award, presented by Jean Lodge, president of the MSA 2015–2016, at the MSA meeting, July 29, 2015, in Edmonton, Ontario, Canada. https://www.youtube.com/watch?v=Zm1c86gRX60.
71. Wasson, R. G., Wasson, V. P. 1957. *Mushrooms, Russia, and History.* Pantheon Books, New York.
72. Gerber, K., Flores, I. G., Ruiz, A. C., Ali, I., et al. 2021. Ethical concerns about psilocybin intellectual property. *ACS Pharmacology & Translational Science* 4(2):573–577. doi.org/10.1021/acsptsci.0c00171.
73. Spiers, N., Labate, B. C., Ermakova, A. O., Farrell, P., et al. 2024. Indigenous psilocybin mushroom practices: an annotated bibliography. *Journal of Psychedelic Studies* 8(1):3–25. doi.org/10.1556/2054.2023.00297.
74. Kneebone, L. R. 1960. Methods for the production of certain hallucinogenic agarics. *Developments in Industrial Microbiology* 1:109. doi.org/10.1007/978-1-4899-5073-4_16.
75. Kilmer, B., et al. 2024. considering alternatives to psychedelic drug prohibition. *RAND Corporation.* doi.org/10.7249/RRA2825-1.
76. Turner, N. J., Cuerrier, A. 2022. "Frog's umbrella" and "ghost's face powder": the cultural roles of mushrooms and other fungi for Canadian Indigenous Peoples. *Botany* 100(2):183–205. doi.org/10.1139/cjb-2021-0052.
77. McCawley, E. L, Brummett, R. E., Dana, G. W. 1962. Convulsions from *Psilocybe* mushroom poisoning. *Proceedings of the Western Pharmacology Society* 5:27–33. PMID: 13932070.
78. Kroeger, P., et al. 2012. *The Outer Spores: Mushrooms of Haida Gwaii.* Vancouver Mycological Society and Mycologue Publications, Sidney-by-the-Sea, British Columbia.
79. Weil, A. T. 1977. *The Use of Psychoactive Mushrooms in the Pacific Northwest: An Ethnopharmacologic Report.* Botanical Museum Leaflets, Harvard University 25(5):131–149. www.jstor.org/stable/41762782
80. Griffiths R. R., Richards W. A., McCann U., Jesse R. 2006. Psilocybin can occasion mystical-type experiences having substantial and sustained personal meaning and spiritual significance. *Psychopharmacology* (Berl). 187(3):268-83; discussion 284-92. doi: 10.1007/s00213-006-0457-5.; Davis A. K., Barrett, F. S., May, D. G., et al. 2021. Effects of Psilocybin-Assisted Therapy on Major Depressive Disorder: A Randomized Clinical Trial. *JAMA Psychiatry*78(5):481–489. doi:10.1001/jamapsychiatry.2020.3285.

Chapter 2. Where to Find Psilocybin Mushrooms

1. Dennis, R. W. G., Wakefield, E. M. 1946. New or interesting British fungi. *Transactions of the British Mycological Society* 29(3):141–166. doi.org/10.1016/S0007-1536(46)80038-X.
2. Wildlife Conservation Society. 2004. Woodpeckers carry fungus in beaks that promotes tree decay. *ScienceDaily* 12 February. www.sciencedaily.com/releases/2004/02/040212090015.htm
3. Jusino, M. A., Lindner, D. L., Banik, M. T., et al. 2016. Experimental evidence of a symbiosis between red-cockaded woodpeckers and fungi. *Proceedings of the Royal Society of Biological Sciences* 283(1827):20160106. doi.org/10.1098/rspb.2016.0106.
4. Rosero, S. 2024. How a trial about the rights of nature led to the discovery of a hallucinogenic mushroom in Ecuador. *El País*, Feb. 10. https://english.elpais.com/science-tech/2024-02-10/how-a-trial-about-the-rights-of-nature-led-to-the-discovery-of-a-hallucinogenic-mushroom-in-ecuador.html
5. Dentinger, B., Furci, G. 2023. *Psilocybe stametsii* Dentinger & Furci, sp. nov. *Index Fungorum* 529.
6. San Ysidro or San Isidro, derived from the Greek *Isidoros*, which means "gifts of Isis."

Chapter 3. How to Identify Psilocybin Mushrooms

1. Psilocybin has been detected in *Massospora levispora* and *Massospora platypediae*, fungal parasites of cicadas that alter their behavior. Boyce, G. R., Gluck-Thaler, E., Slot, J. C., et al. 2019. Psychoactive plant- and mushroom-associated alkaloids from two behavior modifying cicada pathogens. *Fungal Ecology* 41:147–164. doi.org/10.1016/j.funeco.2019.06.002.
2. Stamets, P., Chilton, J. S. 1983. *The Mushroom Cultivator*. Agarikon Press, Olympia, Washington: 204–209.
3. Stamets, P. 2000. *Growing Gourmet and Medicinal Mushrooms*. Ten Speed Press, Berkeley.
4. Kryskow, P., Stamets, P., La Torre, J., et al. 2024. "The mushroom was more alive and vibrant": Patient reports of synthetic versus organic forms of psilocybin. *Journal of Psychedelic Studies* doi.org/10.1556/2054.2024.00379.
5. Edwards, L. 2010. Lightning really does make mushrooms multiply. https://phys.org/news/2010-04-lightning-mushrooms.html.
6. Kobayashi, C., Hiromi, M., and Takuma, T. 2023. Vibrations and mushrooms: Do environmental vibrations promote fungal growth and fruit body formation? *Ecology* 104(6): e4048. doi.org/10.1002/ecy.4048.
7. Gül Dölen and her lab at the Department of Neurology, Johns Hopkins University School of Medicine, demonstrated that during psilocybin journey, a critical window reopens that allows for behavior modification independent with binding to 5HT2A receptors. They demonstrated psilocybin stimulated dendritic spine formation (new neuronal outgrowths that help joins synapses, also known as synaptogenesis), and that psilocybin affects many neural receptors besides 5HT2A. Nardou, R., Sawyer, E., Song, Y. J., et al. 2023. Psychedelics reopen the social reward learning critical period. *Nature* 618:790–798. doi.org/10.1038/s41586-023-06204-3.
8. Lenz, C., Wick, J., Braga, D., et al. 2020. Injury-triggered blueing reactions of *Psilocybe* "magic" mushrooms. *Angewandte Chemie International Edition* 59(4):1450–1454. doi.org/10.1002/anie.201910175.

Chapter 4. How to Create a Psilocybin Mushroom Patch

1. Brown, P. 2011. Magic mushrooms thrive as weeds wane. *The Guardian*, Jan. 3.
2. Although the timing of fruiting is generalized here for fall fruiting, note that *Psilocybe ovoideocystidiata* and edible mushrooms like the Garden Giant, *Stropharia rugoso-annulata*, can fruit in the spring.
3. Hebling, A., Horner, J. W. E., Lehrer, S. B. 1993. Comparison of *Psilocybe cubensis* spore and mycelium allergens. *Journal of Allergy and Clinical Immunology* 91(5):1059–1066. doi.org/10.1016/0091-6749(93)90220-A.
4. Bradshaw, A. J., Backman, T. A., Ramírez-Cruz, V., et al. 2022. DNA authentication and chemical analysis of *Psilocybe* mushrooms reveal widespread misdeterminations in fungaria and inconsistencies in metabolites. *Applied and Environmental Microbiology* 88:e01498-22. doi.org/10.1128/aem.01498-22.
5. Gotvaldová, K., Hájková, K., Borovička, J., et al. 2021. Stability of psilocybin and its four analogs in the biomass of the psychotropic mushroom *Psilocybe cubensis*. *Drug Test Anal.* Feb;13(2):439-446. doi: 10.1002/dta.2950.

Chapter 5. Psilocybin Is Not for Everyone

1. Goff, R., Smith, M., Islam, S., et al. 2024. Determination of psilocybin and psilocin content in multiple *Psilocybe cubensis* mushroom strains using liquid chromatography–tandem mass spectrometry. *Analytica Chimica Acta* 1288:342161. doi.org/10.1016/j.aca.2023.342161.
2. Griffiths, R. R., Johnson., M. W., Richards, W. A., et al. 2011. Psilocybin occasioned mystical-type experiences: immediate and persisting dose-related effects *Psychopharmacology* 218:649–665. doi.org/10.1007/s00213-011-2358-5.
3. Griffiths, R. R., Richards, W. A., Johnson, M. W., et al. 2008. Mystical-type experiences occasioned by psilocybin mediate the attribution of personal meaning and spiritual significance 14 months later. *Journal of Psychopharmacology* 22:621–632.
4. Edinoff, A. N., Swinford, C. R., Odisho, A. S., et al. 2022. Clinically relevant drug interactions with monoamine oxidase inhibitors. *Health Psychology Research* 10(4):39576. doi.org/10.52965/001c.39576.
5. Erritzoe, David et al., 2024. Effect of psilocybin versus escitalopram on depression symptom severity in patients with moderate-to-severe major depressive disorder: observational 6-month follow-up of a phase 2, double-blind, randomised, controlled trial. The Lancet ClinicalMedicine, Volume 0, Issue 0, 102799.
6. Chen, Y., Tian, P., Wang, Z., et al. 2022. Indole acetic acid exerts anti-depressive effects on an animal model of chronic mild stress. *Nutrients*. 14(23):5019. doi: 10.3390/nu14235019.
7. Jabłońska-Ryś, E., Sławińska, A., Stachniuk, A., et al. 2020. Determination of biogenic amines in processed and unprocessed mushrooms from the Polish market. *Journal of Food Composition and Analysis* 92:103492. doi.org/10.1016/j.jfca.2020.103492.

8. Barnett, B. S., Koons, C. J., Van den Eynde, V., et al. 2024. Hypertensive emergency secondary to combining psilocybin mushrooms, extended release dextroamphetamine-amphetamine, and tranylcypromine. *J Psychoactive Drugs* 1–7. doi:10.1080/02791072.2024.2368617.
9. Nayak, S. M., Jackson, H., Sepeda, N. D., et al. 2023. Naturalistic psilocybin use is associated with persisting improvements in mental health and wellbeing: results from a prospective, longitudinal survey. *Frontiers in Psychiatry* 14:1199642. doi.org/10.3389/fpsyt.2023.1199642.
10. Heim, R., Hofmann, A., Brack, A., et al. 1959. Psilocybin and psilocin and processes for their preparation. 8/20/1959, assigned to Sandoz, A. G. Patent no. 959045948. An earlier patent application (no. 12217) was filed on 2/12/1958 with the Israeli patent office with the same title and similar claims, also on behalf of Sandoz.
11. Stanislav Grof interviews Dr. Albert Hofmann, Esalen Institute, Big Sur, California, 1984. https://maps.org/news-letters/v11n2/11222gro.html

Chapter 6. Dosing

1. Rootman, J. M., Kryskow, P., Harvey, K., et al. 2021. Adults who microdose psychedelics report health related motivations and lower levels of anxiety and depression compared to non-microdosers. *Scientific Reports* 11: 22479. doi.org/10.1038/s41598-021-01811-4.
2. Rootman, J. M., Kryskow, P., Harvey, K., et al. 2021. Adults who microdose psychedelics report health related motivations and lower levels of anxiety and depression compared to non-microdosers. *Scientific Reports* 11: 22479. doi.org/10.1038/s41598-021-01811-4.
3. Weiss, B., Roseman, L., Giribaldi, B. et al. 2024. Unique Psychological Mechanisms Underlying Psilocybin Therapy Versus Escitalopram Treatment in the Treatment of Major Depressive Disorder. *Int J Ment Health Addiction* 22, 806–841. doi.org/10.1007/s11469-024-01253-9.
4. Weiss, B., Ginige, I., Shannon, L., et al. 2024. Personality change in a trial of psilocybin therapy v. escitalopram treatment for depression. *Psychological Medicine* 54(1):178–192. doi.org/10.1017/S0033291723001514.
5. Araújo, A. M., Carvalho, F., de Lourdes Bastos, M., et al. 2015. The hallucinogenic world of tryptamines: an updated review. *Archives of Toxicology* 89(8):1151–1173. doi:10.1007/s00204-015-1513-x.
6. Vargas, M. V., Dunlap, L. E., Dong, C., et al. 2023. Psychedelics promote neuroplasticity through the activation of intracellular 5-HT2A receptors. *Science* 379(6633):700–706. doi.org/10.1126/science.adf0435.
7. Sherwood, A. M., Halberstadt, A. L., Klein, A. K., et al. 2020. Synthesis and biological evaluation of tryptamines found in hallucinogenic mushrooms: norbaeocystin, baeocystin, norpsilocin, and aeruginascin. *Journal of Natural Products* 83(2):461–467. doi.org/10.1021/acs.jnatprod.9b01061.
8. Tobin, M. K., Musaraca, K., Disouky, A., et al. 2019. Human hippocampal neurogenesis persists in aged adults and Alzheimer's disease patients. *Cell Stem Cell* 24(6):974–982.e3. doi.org/10.1016/j.stem.2019.05.003.
9. Madsen, M. K., Fisher, P. M., Burmester, D., et al. 2019. Psychedelic effects of psilocybin correlate with serotonin 2A receptor occupancy and plasma psilocin levels. *Neuropsychopharmacology* 44:1328–1334.
10. Rootman, J. M., Kiraga, M., Kryskow, P., et al. 2022. Psilocybin microdosers demonstrate greater observed improvements in mood and mental health at one month relative to non-microdosing controls. *Scientific Reports* 12:11091. doi.org/10.1038/s41598-022-14512-3.
11. Casarotto, P. C., Girych, M., Fred, S. M., et al. 2021. Antidepressant drugs act by directly binding to TRKB neurotrophin receptors. *Cell* 184(5):1299–1313.e19. doi.org/10.1016/j.cell.2021.01.034.
12. Moliner, R., Girych, M., Brunello, C. A., et al. 2023. Psychedelics promote plasticity by directly binding to BDNF receptor TrkB. *Nature Neuroscience* 26:1032–1041. doi.org/10.1038/s41593-023-01316-5.
13. Gasperi, V., Sibilano, M., Savini, I., et al. 2019. Niacin in the central nervous system: an update of biological aspects and clinical applications. *International Journal of Molecular Sciences* 20(4):974. doi.org/10.3390/ijms20040974. PMID: 30813414; PMCID: PMC6412771.
14. Morris, M. C., Evans, D. A., Bienias, J. L., et al. 2004. Dietary niacin and the risk of incident Alzheimer's disease and of cognitive decline. *Journal of Neurology, Neurosurgery & Psychiatry* 75:1093–1099.
15. Kawagishi, H., Shimada, A., Shirai, R., et al. 1994. Erinacines A, B and C, strong stimulators of nerve growth factor (NGF)-synthesis, from the mycelia of *Hericium erinaceum*. *Tetrahedron Letters* 35(10):1569–1572. doi.org/10.1016/S0040-4039(00)76760-8.
16. Eu, W. Z., Chen, Y. J., Chen, W. T., et al. 2021. The effect of nerve growth factor on supporting spatial memory depends upon hippocampal cholinergic innervation. *Translational Psychiatry* 11(1):162. doi.org/0.1038/s41398-021-01280-3.
17. Saitsu, Y., Nishide, A., Kikushima, K., et al. 2019. Improvement of cognitive functions by oral intake of *Hericium erinaceus*. *Biomedical Research* 40(4):125–131. doi.org/10.2220/biomedres.40.125.

18. Nagano, M., Shimizu, K., Kondo, R., et al. 2010. Reduction of depression and anxiety by 4 weeks *Hericium erinaceus* intake. *Biomedical Research* 31(4):231–237. doi.org/10.2220/biomedres.31.231.
19. Yang, P. P., Lin, C.-Y., Lin, T.-Y., et al. 2020. *Hericium erinaceus* mycelium exerts neuroprotective effect in Parkinson's disease: in vitro and in vivo models. *Journal of Drug Research and Development* 6(1). doi.org/10.16966/2470-1009.150.
20. de Veen, B. T. H., Schellekens, A. F. A., Verheij, M. M. M., et al. 2017. Psilocybin for treating substance use disorders? *Expert Review of Neurotherapeutics* 17(2):203–212. doi.org/10.1080/14737175.2016.1220834.
21. Pallav, K., Dowd, S., Villafuerte, J., et al. 2014. Effects of polysaccharopeptide from *Trametes versicolor* and amoxicillin on the gut microbiome of healthy volunteers. *Gut Microbes* 5(4):458–467. doi.org/10.4161/gmic.29558.
22. Nardou, R., Sawyer, E., Song, Y. J., et al. 2023. Psychedelics reopen the social reward learning critical period. *Nature* 618:790–798. doi.org/10.1038/s41586-023-06204-3.
23. Weiss, B., Roseman, L., Giribaldi, B. et al. 2024. Unique Psychological Mechanisms Underlying Psilocybin Therapy Versus Escitalopram Treatment in the Treatment of Major Depressive Disorder. *Int J Ment Health Addiction* 22, 806–841 doi.org/10.1007/s11469-024-01253-9.
24. Siegel, J. S., Subramanian, S., Perry, D., et al. 2024. Psilocybin desynchronizes the human brain. *Nature* 632:131–138. doi.org/10.1038/s41586-024-07624-5
25. Pollan, M. 2019. *How to Change Your Mind.* Penguin Books, New York.
26. Petri G., Expert P., Turkheimer F., Carhart-Harris R., et al. 2014. Homological scaffolds of brain functional networks. *J. R. Soc. Interface* 11:20140873. doi.org/10.1098/rsif.2014.0873.
27. Readers may benefit from seeing the chapter "Good Tips for Great Trips" in my 1996 book *Psilocybin Mushrooms of the World* (Ten Speed Press, Berkeley).
28. Johnson, M. W., Garcia-Romeu, A., Griffiths, R. R. 2017. Long-term follow-up of psilocybin-facilitated smoking cessation. *American Journal of Drug and Alcohol Abuse* 43(1):55–60. doi.org/10.3109/00952990.2016.1170135. Erratum in: *American Journal of Drug and Alcohol Abuse* 43(1):127. PMID: 27441452.
29. Bogenschutz, M. P., Ross, S., Bhatt, S., et al. 2022. Percentage of heavy drinking days following psilocybin-assisted psychotherapy vs placebo in the treatment of adult patients with alcohol use disorder: a randomized clinical trial. *JAMA Psychiatry* 79(10):953–962. doi.org/10.1001/jamapsychiatry.2022.2096.
30. Breakthrough therapy. Available via https://www.fda.gov/patients/fast-track-breakthrough-therapy-accelerated-approval-priority-review/breakthrough-therapy. See also Saplakoglu, Y. 2019. FDA calls psychedelic psilocybin a "breakthrough therapy" for severe depression. Available via https://www.livescience.com/psilocybin-depression-breakthrough-therapy.html
31. Griffiths, R. R., Johnson, M. W., Carducci, M. A., et al. 2016. Psilocybin produces substantial and sustained decreases in depression and anxiety in patients with life-threatening cancer: a randomized double-blind trial. *Journal of Psychopharmacology* 30(12):1181–1197. doi.org/10.1177/0269881116675513.
32. Nayak, S. M., Bari, B. A., Yaden, D. B., et al. 2023. A Bayesian reanalysis of a trial of psilocybin versus escitalopram for depression. *Psychedelic Medicine* 1(1):18–26. doi.org/10.1089/psymed.2022.0002.
33. Weiss, B., Ginige, I., Shannon, L., et al. 2023. Personality change in a trial of psilocybin therapy v. escitalopram treatment for depression. *Psychological Medicine* 54(1):178–192. doi.org/10.1017/S0033291723001514.
34. Gukasyan, N., Davis, A. K., Barrett, F. S., et al. 2022. Efficacy and safety of psilocybin-assisted treatment for major depressive disorder: prospective 12-month follow-up. *Journal of Psychopharmacology* 36(2):151–158. doi.org/10.1177/02698811211073759.
35. Goodwin, G. M., Aaronson, S. T., Alvarez, O., et al. 2022. Single-dose psilocybin for a treatment-resistant episode of major depression. *New England Journal of Medicine* 387(18):1637–1648. doi.org/10.1056/NEJMoa2206443.
36. Kettner, H., Gandy, S., Haijen, E. C. H. M., et al. 2019. From egoism to ecoism: psychedelics increase nature relatedness in a state-mediated and context-dependent manner. *International Journal of Environmental Research and Public Health* 16(24):5147. doi.org/10.3390/ijerph16245147.
37. Forstmann, M., Kettner, H. S., Sagioglou, C., et al. 2023. Among psychedelic-experienced users, only past use of psilocybin reliably predicts nature relatedness. *Journal of Psychopharmacology* 37(1):93–106. doi.org/10.1177/02698811221146356.
38. Forstmann, M., 2023. *Journal of Psychopharmacology* 37(1):93–106. doi.org/10.1177/02698811221146356.
39. Hendricks, P. S., Crawford, M. S., Cropsey, K. L., et al. 2018. The relationships of classic psychedelic use with criminal behavior in the United States adult population. *Journal of Psychopharmacology* 32(1):37–48. doi.org/10.1177/0269881117735685.
40. Thiessen, M. S., Walsh, Z., Bird, B. M., et al. 2018. Psychedelic use and intimate partner violence: the role of emotion regulation. *Journal of Psychopharmacology*

32(7):749–755. doi.org/10.1177/0269881118771782.

41. Jones, G., Ricard, J. A., Lipson, J., Nock, M. K. 2022. Associations between classic psychedelics and opioid use disorder in a nationally representative U.S. adult sample. *Scientific Reports* 12(1):4099. doi:10.1038/s41598-022-08085-4.

Chapter 8. The Psilocybin-Active Species

1. Gotvaldová, K., Borovička, J., Hájková, K., et al. 2022. Extensive collection of psychotropic mushrooms with determination of their tryptamine alkaloids. *International Journal of Molecular Sciences* 23(22):14068. doi.org/10.3390/ijms232214068.
2. Besl, H. 1993. *Galerina steglichii* spec. nov., ein halluzinogener Häubling. *Zeitschrift für Mykologie* 59(2):215–218.
3. Guzmán, G. 1995. Supplement to the monograph of the genus *Psilocybe*. Taxonomic monographs of agaricales. *Bibliotheca Mycologica* 159:91–141. www.researchgate.net/publication/254725842
4. Bradshaw, A. J., Ramírez-Cruz, V., Awan, A. R., Dentinger, B. T. M. 2024. Phylogenomics of the psychoactive mushroom genus *Psilocybe* and evolution of the psilocybin biosynthetic gene cluster. *Proceedings of the National Academy of Sciences* 121(3):1–9. doi.org/10.1073/pnas.2311245121.
5. Fries, E. 1821. Systema mycologicum: sistens fungorum ordines, genera et species, huc usque cognitas, quas and normam methodi naturalis determinavit. *Ex Officina Berlingiana,* Lund, Sweden: 289. doi.org/10.5962/bhl.title.5378.
6. Cho, H. J., Lee, H., Park, J. Y., et al. 2016. Seven new recorded species in five genera of the Strophariaceae in Korea. *Mycobiology* 44(3):137–145. doi.org/10.5941/MYCO.2016.44.3.137.
7. Earle, F. S. 1906. Algunos hongos cubanos. *Información Anual Estación Central Agronomica Cuba* 1:225–242.
8. Singer, R. 1958. Mycological investigations on teonanácatl, the Mexican hallucinogenic mushroom. Part I. The history of teonanácatl, field work and culture work. *Mycologia* 50:239–261.
9. Guzmán, G. 1984. *The Genus* Psilocybe: *A Systematic Revision of the Known Species Including the History, Distribution and Chemistry of the Hallucinogenic Species.* J. Cramer Verlag, Liechtenstein.
10. Noordeloos, M. E. 1995. Strophariaceae Sing. & Smith. *Flora Agaricina Neerlandica* 4:81.
11. Singer, R., Smith, A. H. 1946. The taxonomic position of *Pholiota mutabilis* and related species. *Mycologia* 38(5):500–523. doi.org/10.1080/00275514.1946.12024074.
12. Noordeloos, M. E. 2011. Strophariaceae s. l. *Fungi Europaei* 13. Edizioni Candusso, Alassio, Italy.
13. Redhead, S. A., Moncalvo, J.-M., Vilgalys, R., et al. 2007. Proposal to conserve the name *Psilocybe* (Basidiomycota) with a conserved type. *Taxon* 56(1):255–257. www.jstor.org/stable/25065762.
14. McNeill, J., Turland, N. J., Monro, A., et al. 2011. XVIII International Botanical Congress: preliminary mail vote and report of Congress action on nomenclature proposals. *Taxon* 60(5):1507–1520. www.jstor.org/stable/41317562.
15. Borovička, J., Oborník, M., Stříbrný, J., et al. 2015. Phylogenetic and chemical studies in the potential psychotropic species complex of *Psilocybe atrobrunnea* with taxonomic and nomenclatural notes. *Persoonia* 34:1–9. doi.org/10.3767/003158515X685283.
16. Bradshaw et al. 2024. *Proceedings of the National Academy of Sciences* 121(3):1– 9. doi. org/10.1073/pnas.2311245121.
17. Favre, J. 1939. *Psilocybe turficola.* In Gargominy O. 2022. TAXREF. Version 4.9. UMS PatriNat (OFB-CNRS-MNHN), Paris. Checklist dataset doi.org/10.15468/vqueam.
18. Borovička, J., Oborník, M., Stříbrný, J., et al. 2015. Phylogenetic and chemical studies in the potential psychotropic species complex of *Psilocybe atrobrunnea* with taxonomic and nomenclatural notes. *Persoonia* 34:1–9. doi.org/10.3767/003158515X685283.
19. Dentinger, B. 2023. Personal communication with the author.
20. Miller, D. Danny's DNA Discoveries: *Psilocybe* of the PNW. https://www.alpental.com/psms/ddd/Hymenogastraceae/Psilocybe.htm
21. Stamets, P., Gartz, J. 1995. A new caerulescent *Psilocybe* from the Pacific Coast of Northwestern North America. *Integration: Journal for Mind Moving Plants and Culture* 6:21–27. [MB#446057]
22. Sherwood, A. M., Halberstadt, A. L., Klein, A. K., et al. 2020. Synthesis and biological evaluation of tryptamines found in hallucinogenic mushrooms: norbaeocystin, baeocystin, norpsilocin, and aeruginascin. *Journal of Natural Products* 83:461–467. doi.org/10.1021/acs.jnatprod.9b01061.
23. Chadeayne, A. R., Pham, D. N. K., Reid, B. G., et al. 2020. Active metabolite of aeruginascin (4-hydroxy-N,N,N-trimethyltryptamine): synthesis, structure, and serotonergic binding affinity. *ACS Omega* 5(27):16940–16943. doi.org/10.1021/acsomega.0c02208.
24. Gartz, J. 1989. Analysis of aeruginascin in fruit bodies of the mushroom *Inocybe aeruginascens. International*

Journal of Crude Drug Research 27:141–144. doi.org/10.3109/13880208909053954.

25. Jensen, N., Gartz, J., Laatsch, H. 2006. Aeruginascin, a trimethylammonium analogue of psilocybin from the hallucinogenic mushroom *Inocybe aeruginascens*. *Planta Medica* 72:665–666.
26. Beug, M., Bigwood, J. 1981. Quantitative analysis of psilocybin and psilocin in *Psilocybe baeocystis* Singer & Smith by high-performance liquid chromatography and by thin-layer chromatography. *Journal of Chromatography* 207:370–385.
27. Beug, M., Bigwood, J. 1982. Psilocybin and psilocin levels in twenty species from seven genera of wild mushrooms in the Pacific Northwest (U.S.A.). *Journal of Ethnopharmacology* 5:271–278.
28. Canan, J., Ostuni S., Rockefeller. A., Birkebak, J. 2024. *Psilocybe caeruleorhiza*: a new, cold weather fruiting species of psilocybin containing mushroom from the Midwest in section *Aztecorum*. *McIlvainea* 33:3 [MB#853101]
29. Murrill, W. A., 1923. Dark spored agarics 5. *Psilocybe*. *Mycologia* 15:1–55.
30. Heim R. Wasson, R. G. 1958. Les champignons hallucinogenes du mexique. *Museum de historie naturelle* 268–71, Paris.
31. Bradshaw et al. 2024. *Proceedings of the National Academy of Sciences* 121(3):1– 9. doi. org/10.1073/pnas.2311245121.
32. Dentinger, B. 2023. Personal communication with the author.
33. Singer, R., Smith, A. H. 1958. New species of *Psilocybe*. *Mycologia* 50:141–142.
34. Bradshaw et al. 2024. *Proceedings of the National Academy of Sciences* 121(3):1– 9. doi. org/10.1073/pnas.2311245121.
35. Goff, R., Smith, M., Islam, S., et al. 2024. Determination of psilocybin and psilocin content in multiple *Psilocybe cubensis* mushroom strains using liquid chromatography—tandem mass spectrometry. *Analytica Chimica Acta* 1288:342161. doi.org/10.1016/j.aca.2023.342161.
36. Gotvaldová, K., Hájková, K., Borovička, J., et al. 2021. Stability of psilocybin and its four analogs in the biomass of the psychotropic mushroom *Psilocybe cubensis*. *Drug Testing and Analysis* 13(2):439–446. doi.org/10.1002/dta.2950.
37. Stamets, P., Chilton, J. S. 1983. *The Mushroom Cultivator*. Agarikon Press, Olympia, Washington.
38. Stamets, P. 2000. *Growing Gourmet and Medicinal Mushrooms*. Ten Speed Press, Berkeley.
39. McTaggart, A.R., McLaughlin, S., Slot, J.C., et al. 2023. Domestication through clandestine cultivation constrained genetic diversity in magic mushrooms relative to naturalized populations. *Current Biology* 33(23):5147-5159. doi.org/10.1016/j.cub.2023.10.059
40. Helbling, A., Horner, W. E., Lehrer, S. B. 1993. Comparison of *Psilocybe cubensis* spore and mycelium allergens. *Journal of Allergy and Clinical Immunology* 91(5):1059–1066. doi.org/10.1016/0091-6749(93)90220-a.
41. Beug, M., Bigwood, J. *Journal of Ethnopharmacology* 5:271–278.
42. Gartz, J. 1994. Extraction and analysis of indole derivatives from fungal biomass. *Journal of Basic Microbiology* 34(1):17–22. doi.org/10.1002/jobm.3620340104.
43. Stijve, T., Kuyper, T. W. 1985. Occurrence of psilocybin in various higher fungi from several European countries. *Planta Medica* 51(5):385–387. doi.org/10.1055/s-2007-969526.
44. Singer, R., Smith, A. H. 1958. Mycological investigations on teonanácatl, the Mexican hallucinogenic mushroom. Part II. A taxonomic monograph of *Psilocybe*, section Caerulescentes. *Mycologia* 50(2):262–303. doi.org/10.1080/00275514.1958.12024726.
45. Stamets, P. E., Beug, M. W., Bigwood, J. E., Guzmán, G. 1980. A new species and new variety of *Psilocybe* from North America. *Mycotaxon* 11(2):476–484.
46. Guzmán, G. 2000. New species and new records of *Psilocybe* from Spain, the U.S.A. and Mexico, and a new case of poisoning by *Psilocybe herrerae*. *Documents Mycologiques* 29(116):41–52.
47. Gotvaldová, K., Borovička, J., Hájková, K., et al. 2022. Extensive collection of psychotropic mushrooms with determination of their tryptamine alkaloids. *Int J Mol Sci.* Nov 15;23(22):14068. doi:10.3390/ijms232214068.
48. Fernández-Sasia, R. 2006. *Psilocybe hispanica* Guzmán, un taxón novedoso en nuestro entorno. *Errotari* 3:73–76.
49. Akers, B., Piper, A., Ruiz, J. F., Ruck, C. 2011. A prehistoric mural in Spain depicting neurotropic *Psilocybe* mushrooms. *Economic Botany* 65(2):121–128. doi.org/10.1007/s12231-011-9152-5.
50. Guzmán, G. 1984. *The Genus* Psilocybe: *A Systematic Revision of the Known Species Including the History, Distribution and Chemistry of the Hallucinogenic Species.* J. Cramer Verlag, Liechtenstein.
51. Guzmán, G., Greene, J., Ramirez-Guillén, F. 2010. A new for science neurotropic species of *Psilocybe* (Fr.) P. Kumm. (Agaricomycetideae) from the western United States. *International Journal of Medicinal Mushrooms* 12(2):201–204. doi.org/10.1615/IntJMedMushr.v12.i2.110.

52. van der Merwe, B., Rockefeller, A., Kilian, A., et al. 2024. A description of two novel *Psilocybe* species from southern Africa and some notes on African traditional hallucinogenic mushroom use. *Mycologia* 116(5): 821–834.

53. Guzmán, G. 1983. *The Genus Psilocybe*, J. Cramer, Vaduz, Germany.

54. Stamets et al. 1980. *Mycotaxon* 11:476-484.

55. Stijve, T., Kuyper, T. W. 1985. Occurrence of psilocybin in various higher fungi from several European countries. *Planta Medica* 51(5):385–387. doi.org/10.1055/s-2007-969526.

56. Beug, M., Bigwood, J., 1982. *Journal of Ethnopharmacology* 5:271–278.

57. Borovička, J., Oborník, M., Stříbrný, J., et al. 2015. Phylogenetic and chemical studies in the potential psychotropic species complex of *Psilocybe atrobrunnea* with taxonomic and nomenclatural notes. *Persoonia* 34:1–9. doi.org/10.3767/003158515X685283.

58. Guzmán, G., Walstad, L., Gándara, E., et al. 2007. A new bluing species of *Psilocybe*, section *Stuntzii*, from New Mexico, USA. *Mycotaxon* 99:223–226.

59. Stamets, P., Chilton, J. 1983. *The Mushroom Cultivator*. Agarikon Press, Olympia, Washington.

60. Gartz, J., Reid, D., Eicker, A., Smith, M. T. 1995. *Psilocybe natalensis* sp. nov.: the first indigenous bluing member of the Agaricales of South Africa. *Integration* 6:29–34.

61. Gartz, J. (Taake, C., trans. & ed.) 1996. *Magic Mushrooms around the World: A Scientific Journey across Cultures and Time—The Case for Challenging Research and Value Systems*. LIS Publications, Los Angeles.

62. Ostuni, S., Rockefeller, A., Jacobs, J., Birkebak, J. 2024. *Psilocybe niveotropicalis*: a new species of psilocybin containing mushroom from South Florida. *McIlvainea* 33. https://namyco.org/wp-content/uploads/2024/03/NewSpeciesofwood-lovingPsilocybe.pdf

63. Dentinger, B. 2024. Personal communication with the author.

64. Gartz, J. 1994. Extraction and analysis of indole derivatives from fungal biomass. *Journal of Basic Microbiology* 34(1):17–22. doi.org/10.1002/jobm.3620340104.

65. Stijve, T., de Meijer, A. A. R. 1993. Macromycetes from the State of Parana, Brazil. 4. The psychoactive species. *Brazilian Archives of Biology and Technology* 36(2):313–329.

66. Guzmán, G., Bandala, V. M., Allen, J. W. 1993. A new bluing *Psilocybe* from Thailand. *Mycotaxon* XLVI:155–160.

67. Keay, S. M., Brown, A. E. 1990. Colonization by *Psilocybe semilanceata* of roots of grassland flora. *Mycological Research* 94(1):49–56. doi.org/10.1016/S0953-7562(09)81263-X.

68. Ohenoja, E., Jokiranta, J., Mäkinen, T., et al. 1987. The occurrence of psilocybin and psilocin in Finnish fungi. *Journal of Natural Products.* 50(4): 741–44. doi:10.1021/np50052a030.

69. Gartz J. 1994. New aspects of the occurrence, chemistry and cultivation of European hallucinogenic mushrooms. *Annali del Museo Civico di Rovereto* 8:107–23.

70. Stijve, T., Kuyper, T. W. 1985. Occurrence of psilocybin in various higher fungi from several European countries. *Planta Medica* 51(5):385–387. doi.org/10.1055/s-2007-969526.

71. Christiansen, A. L., Rasmussen, K. E., Hoiland, K. 1981. The content of psilocybin in Norwegian *Psilocybe semilanceata*. *Planta Medica* 42:229–235.

72. Ohenoja, E., Jokiranta, J., Makinen, T., et al. 1987. The occurrence of psilocybin and psilocin in Finnish fungi. *Journal of Natural Products* 50(4):741–744.

73. Bradshaw, A. J. 2022.

74. Keay, S. M., Brown, A. E. 1990. Colonization by *Psilocybe semilanceata* of roots of grassland flora. *Fungal Biology* 94; 49–56.

75. Gotvaldová, K., Borovička, J., Hájková, K., et al. 2022. Extensive collection of psychotropic mushrooms with determination of their tryptamine alkaloids. *International Journal of Molecular Sciences* 23(22):14068. doi.org/10.3390/ijms232214068.

76. Guzmán, G. 1995. Supplement to the monograph of the genus *Psilocybe*. Taxonomic monographs of agaricales. *Bibliotheca Mycologica* 159:91–141.

77. Beug M. 2011. The genus *Psilocybe* in North America. *Fungi Magazine*. 4(3):6–17.

78. Musshoff, F., Madea, B., Beike, J. 2000. Hallucinogenic mushrooms on the German market—simple instructions for examination and identification. *Forensic Science International*. 113(1–3): 389–95. doi:10.1016/S0379-0738(00)00211-5.

79. Chang, Y. S., Gates, G. M., Ratkowsky, D. A. 2006. Some new species of the Strophariaceae (Agaricales) in Tasmania. *Australasian Mycologist* 24:53–68.

80. Guzmán, G., Watling, R. 1978. Studies in Australian agarics and boletes. I. Some species of *Psilocybe*. *Notes of the Royal Botanic Garden Edinburgh* 36(1):199–210.

81. See comment ref. 16. *bioRxiv* 12.13.520147. doi.org/10.1101/2022.12.13.520147.

82. Guzmán, G., Tapia, F., Stamets, P. 1997. A new bluing *Psilocybe* from U.S.A. *Mycotaxon* 65:191–196.

83. McTaggart, A., Scarlett, K., Slot, J., et al. 2023. Wood-loving magic mushrooms from Australia are saprotrophic invaders in the northern hemisphere. *Authorea.* doi:10.22541/au.170019739.91075425/v1.

84. Heim, R.,Wasson, R. W. 1958. *Les champignons hallucinogénes du Mexique*. Editions du Museum National d'Histoire Naturelle, Paris.

85. Guzmán, G. 1984. *The Genus* Psilocybe: *A Systematic Revision of the Known Species Including the History, Distribution and Chemistry of the Hallucinogenic Species.* J. Cramer Verlag, Liechtenstein.

86. Strauss, D., Ghosh, S., Murray, Z., Gryzenhout, M. 2023. Global species diversity and distribution of the psychedelic fungal genus *Panaeolus*. *Heliyon* 9(6):e16338. doi.org/10.1016/j.heliyon.2023.e16338.

87. Ola'h, G. M. 1969. *Le genre Panaeolus: Essai taxonimique et physiologique*. Laboratoire de Cryptogamie, Paris.

88. Stijve, T., Kuyper, T. W. 1985. Occurrence of psilocybin in various higher fungi from several European countries. *Planta Medica* 51(5):385–387. doi.org/10.1055/s-2007-969526.

89. Stijve, T., de Meijer, A. A. R. 1993. Macromycetes from the State of Parana, Brazil. 4. The psychoactive species. *Brazilian Archives of Biology and Technology* 36(2):313–329.

90. Gurevich L.S. 1993. Indole derivatives in certain *Panaeolus* species from east Europe and siberia. *Mycol. Res.* 97:251–254. doi:10.1016/S0953-7562(09)80249-9.

91. Stijve, T. 1992. Psilocin, psilocybin, serotonin and urea in *Panaeolus cyanescens* from various origin. *Persoonia* 15(1):117–121. Rijksherbarium, Hortus Botanicus, Leiden.

92. Laussmann, T., Meier-Giebing, S. 2010. Forensic analysis of hallucinogenic mushrooms and khat (*Catha edulis* Forsk) using cation-exchange liquid chromatography. *Forensic Science International* 195(1–3):160–164. doi.org/10.1016/j.forsciint.2009.12.013.

93. Murrill, W. 1923. Dark spored agarics. *Mycologia* 15(1):17.

94. Smith, A. H. 1948. Dark spored agarics. *Mycologia* 40(6):685.

95. Gotvaldová, K., Borovička, J., Hájková, K., et al. 2022. Extensive collection of psychotropic mushrooms with determination of their tryptamine alkaloids. *International Journal of Molecular Sciences* 23(22):14068. doi.org/10.3390/ijms232214068.

96. Fayod, V. F. 1889. Prodrome d'une histoire naturelle des Agaricinés. *Annales des Sciences Naturelles, Botanique.* 9:181–411.

97. Hausknecht, A. 2009. Monograph of the Genera Conocybe Fayod, Pholiotina Fayod in Europe. Italy: Edizioni Candusso, Alassio. 2009.

98. Song, H. B., Bau, T. 2024. Resolving the polyphyletic origins of Pholiotina s.l. (Bolbitiaceae, Agaricales) based on Chinese materials and reliable foreign sequences. *Mycosphere* 15(1), 1595–1674, doi.10.5943/mycosphere/15/1/14.

99. Ohenoja, E., Jokirant, J., Makine, T., et al. 1987. The occurrence of psilocybin and psilocin in Finnish fungi. *Journal of Natural Products* 54(4):741–744.

100. Beug, M., Bigwood, J. *Journal of Ethnopharmacology* 5:271–278.

101. Christiansen, A. L., Rasmussen, K. E., Hoiland, K. 1984. Detection of psilocybin and psilocin in Norwegian species of *Pluteus* and *Conocybe*. *Planta Medica* 45: 341–343.

102. Gartz, J. 1992. New aspects of the occurrence, chemistry and cultivation of European hallucinogenic mushrooms. *Annali Musei Civic di Rovereto* 8:107–124.

103. Halama, M., Poliwoda, A., Jasicka-Misiak, I., et al. 2014. *Pholiotina cyanopus*, a rare fungus producing psychoactive tryptamines. *Open Life Sciences* 10(1):40–51. doi.org/10.1515/biol-2015-0005.

104. Hausknecht, A., Krisai-Greilhuber, I., Voglmayr, H. 2004. Type studies in North American species of Bolbitiaceae belonging to the genera *Conocybe* and *Pholiotina*. *Österreichische Zeitschrift für Pilzkunde* 13:153–235.

105. Guzmán-Dávalos, L., Mueller, G. M., Cifuentes, J., et al. 2003. Traditional infrageneric classification of *Gymnopilus* is not supported by ribosomal DNA sequence data. *Mycologia* 95:1204–1214. doi.org/10.2307/3761920.

106. One interesting bioassay from the United States is worth mentioning. David Arora (in *Mushrooms Demystified*, 1986:410, Ten Speed Press, Berkeley) reports on the fate of one surprised victim who, after ingesting this mushroom and unexpectedly succumbing to its effects, exclaimed on the way to the hospital "If this is the way you die from mushroom poisoning, then I'm all for it."

107. Chaverra-Muñoz, L., Briem, T., Hüttel, S. 2022. Optimization of the production process for the anticancer lead compound illudin M: improving titers in shake-flasks. *Microb Cell Fact.* 21(1):98. doi:10.1186/s12934-022-01827-z.

108. Tanaka, M., Hashimoto, K., Okuno, T., Shirahama, H. 1993. Neurotoxic oligoisoprenoids of the hallucinogenic mushroom, *Gymnopilus spectabilis*. *Phytochemistry* 34(3):661–664. doi.org/10.1016/0031-9422(93)85335-O.

109. Hatfield, G. M., Valdes, L. J., Smith, A. H. 1978. The occurrence of psilocybin in *Gymnopilus* species. *Lloydia* 41:140–144.

110. Lee, I. K., Cho, S. M., Seok, S. J., Yun, B. S. 2008. Chemical constituents of *Gymnopilus spectabilis* and their antioxidant activity. *Mycobiology* 36(1):55–59. doi.org/10.4489/MYCO.2008.36.1.055.

111. Stijve, T., Kuyper, T. W. 1988. Absence of psilocybin in species of fungi previously reported to contain psilocybin and related tryptamine derivatives. *Persoonia* 13:463–465. https://www.repository.naturalis.nl/document/569679

112. Thorn, R. G., Malloch, D. W., Saar, I., et al. 2020. New species in the *Gymnopilus junonius* group (Basidiomycota: Agaricales). *Botany* 98(6):293–315. doi.org/10.1139/cjb-2020-0006.

113. Hesler, L. R. 1969. North American species of *Gymnopilus*. *Mycological Memoirs* no. 3. Hafner Publishers, New York.

114. Hatfield, G. M., Valdes, L. J. 1978. The occurrence of psilocybin in Gymnopilus species. *Lloydia* 41(2):140–4. PMID: 565861.

115. Gartz, J. 1989. Occurrence of psilocybin, psilocin and baeocystin in *Gymnopilus purpuratus*. *Persoonia* 14:19–22.

116. Gartz, J. 1994. Extraction and analysis of indole derivatives from fungal biomass. *Journal of Microbiology* 34:17–22.

117. Tanaka, M., Hashimoto, K., Okuno, T., et al. 1993. Neurotoxic oligiosoprenoids of the hallucinogenic mushroom *Gymnopilius spectabilis*. *Phytochemistry* 34(2):661–664.

118. Stijve, T., Kuyper, T. W. 1985. Occurrence of psilocybin in various higher fungi from several European countries. *Planta Medica* 51(5):385–387. doi.org/10.1055/s-2007-969526.

119. Stijve, T., Klan, J., Kuyper, T. W. 1985. Occurrence of psilocybin and baeocystin in the genus *Inocybe* (Fr.) Fr. *Persoonia* 12(4):469–473.

120. Gotvaldová K., Borovička J., Hájková K., et al. 2022. Extensive Collection of Psychotropic Mushrooms with Determination of Their Tryptamine Alkaloids. *Int J Mol Sci.* Nov 15;23(22):14068. doi:10.3390/ijms232214068.

121. Stijve, T., Klan, J., Kuyper, T. W. 1985.

122. Gartz, J. 1986. Psilocybin in Mycelkulturen von *Inocybe aeruginascens*. *Biochemie und Physiologie der Pflanzen* 181(7):511–517. doi.org/10.1016/S0015-3796(86)80042-7.

123. Stijve, T., Klan, J., Kuyper, T. W. 1985.

124. Cullington, P., Outen, A. 2009. *Inocybe erinaceomorpha*: why it should be regarded as a good species. *Field Mycology* 10(1):11–14. doi.org/10.1016/S1468-1641(10)60490-2.

125. Andersson, C., Gry, J., Kristinsson, J., 2009. Occurrence and use of hallucinogenic mushrooms containing psilocybin alkaloids. *TemaNord* Nordisk Ministerråd.

126. Stijve, T., Kuyper, T. W. 1985. Occurrence of psilocybin in various higher fungi from several European countries. *Planta Medica* 51(5):385–387. doi.org/10.1055/s-2007-969526.

127. Christiansen, A. L., Rasmussen, K. E., Høiland, K. 1984. Detection of psilocybin and psilocin in Norwegian species of *Pluteus* and *Conocybe*. *Planta Medica* 50(4):341–343. doi.org/10.1055/s-2007-969726.

128. Ohenoja, E., Jokiranta, J., Mäkinen, T., et al. 1987. The occurrence of psilocybin and psilocin in Finnish fungi. *Journal of Natural Products* 50(4):741–744. doi.org/10.1021/np50052a030.

129. Stijve, T., Kuyper, T. W. 1985. Occurrence of psilocybin in various higher fungi from several European countries. *Planta Medica* 51(5):385–387. doi.org/10.1055/s-2007-969526.

130. Christiansen, A. L., Rasmussen, K. E., Hoiland, K. 1984.

Resources

A Few of Many Helpful Field Guides

Fungi of Temperate Europe by Jens H. Petersen and Thomas Læssøe (Princeton University Press, 2019)

Mushrooms of British Columbia by Andy MacKinnon and Kem Luther (The Royal British Columbia Museum, 2021)

Mushrooms of Cascadia: A Comprehensive Guide to Fungi of the Pacific Northwest by Noah Siegel and Christian Schwarz (Backcountry Press, 2024)

Mushrooms of Cascadia, Second Edition: An Illustrated Key to the Fungi of the Pacific Northwest by Michael Beug (Ten Speed Press, 2024)

Mushrooms of the Pacific Northwest by Steve Trudell and Joe Ammirati (Timber Press, 2009)

Websites

Mushroom Effects and Toxicology

Note: In the United States, due to Health Insurance Portability and Accountability Act (HIPPA) regulations, very little patient information can be shared with mushroom experts, which limits the ability of mycologists to help with diagnoses and treatment protocols.

Erowid
erowid.org

North American Mycological Association
namyco.org

Mushroom Identification and Information

Note: It is not wise to rely solely on websites for mushroom identification. Be absolutely sure that mushrooms are safe before consuming or giving to others.

Danny's DNA Discoveries
alpental.com/psms/ddd/

Global Biodiversity Information Facility
gbif.org

iNaturalist
inaturalist.org

Index Fungorum
indexfungorum.org

Mushroom Expert
mushroomexpert.com

Mushroom Observer
mushroomobserver.org

MycoBank
mycobank.org

MycoMatch
mycomatch.com

Mushroom Latin Name Pronunciations

Listen to American mycologist Dr. Alexander Smith and European mycologist Dr. Rolf Singer pronounce the Latin names of fungi. Prepared by Kit Scates Barnhart and Kenneth W. Cochran for the Education Committee, North American Mycological Society.

Snohomish County Mycological Society
scmsfungi.org (under the "Mycology" tab)

Psilocybin Churches and Religious Organizations

The Church of Ambrosia, Oakland, California
ambrosia.church

The Divine Assembly, Salt Lake City, Utah
thedivineassembly.org

The Healing Warrior Church, Austin, Texas
healingwarriorfoundation.org

Lotus Entheogenic Church, Oakland, California
lotuschurch.com

Microdosing Information

Have A Good Trip: Exploring The Magic Mushroom Experience by Eugenia Bone (Flatiron Books, 2024)

Microdosing: Health, Healing, and Enhanced Performance by James Fadiman and Jordan Gruber (St. Martin's Essentials, 2025)

Mushroom Supplements

Host Defense:
hostdefense.com

Fungi Perfecti, LLC:
fungi.com

Continuously Updated Summary of Scientific References
mushroomreferences.com

This book is part of a never-ending story. For ongoing updates to *Psilocybin Mushrooms in Their Natural Habitats*, go to psilocybinmushrooms intheirnaturalhabitats.com.

Image Credits

All photographs are by Paul Stamets with the exception of those listed below.

Front cover: *Psilocybe weraroa* (top right) by D. B. Townsend; *Psilocybe hoogshagenii* (vertical image to right of title), *Psilocybe cubensis* (middle left), and *Psilocybe caerulescens* (bottom left) by Alan Rockefeller; *Psilocybe pelliculosa* (middle right) and *Psilocybe baeocystis* (bottom right) by Paul Stamets.

Back cover: by Christian Schwarz (*Psilocybe semilanceata*)

Page 2, 88, and 229 by Pamela Kryskow

Page 6 "The Generous Queen" by Autumn Skye, 30" x 40" Acrylic on Canvas 2024

Page 8, 30, and 48 by Dr. Patrick C. Hickey

Page 10 by John Stamets

Page 18, 22, and 98 by Alex Grey

Page 19, 193, 198, and 202 by Bryn Dentinger

Page 20 by Gary Lincoff

Page 25 by Jonathan Meader

Page 25 and 27 by Jean-Dominique Lajoux

Page 27, 161, 162, and 185 by Ignacio Seral Bozal

Page 27 by Unknown

Page 28 by Dakota Wint

Page 29 by the Eric P. Newman Numismatic Education Society

Page 31 by Unknown

Page 35 by Dusty Yao

Page 35 by Casey Mullen

Page 37 by Unknown

Page 41, 42, and 44 by Allan B. Richardson, Courtesy of The Tina & Gordon Wasson Ethnomycological Collection, Harvard University

Page 41 and 44 by Inti Garcìa Flores

Page 41 by Don Josè Àvila

Page 49 by Azureus Stamets

Page 50 by Bill Stamets

Page 50 and 51 by Chris W. Nelson

Page 51 Courtesy of MAPS

Page 51 Courtesy of Roland Griffiths

Page 54 Courtesy of Bryn Dentinger

Page 56 by Rich Lotus

Page 56 by Jeremy Bigwood

Page 57 and 58 by Cosmo Sheldrake

Page 60 by Lee Stein

Page 70 by Dave Hodges

Page 71 and 170 by Nicolas Schwab

Page 72 and 153 by Kit Scates Barnhart

Page 72, 122, 125, 189, and 205 by Michael W. Beug

Page 92 by Mark Henson

Page 108 by Public Domain

Page 109 by Josh Siegel

Page 111 by Amanda Feilding, Countess of Wemyss

Page 112 by Paul Noth

Page 122, 127, 128, and 129 by Kit Barnhart Scates and Michael W. Beug

Page 124, 125, and 128 by João Silva

Page 126 by Steve Cividanes

Page 129 by Harley Barnhart and Michael W. Beug

Page 129 by Erlon Bailey

Page 130 by Nelson Andres Masis

Page 135, 136, 138, 147, 148, 151, 152, 163, 164, 172, 173, 174, 179, 200, 201, 208, 215, 222, and 227 by Alan Rockefeller

Page 137 by Renèe Lebeuf

Page 145 by Santiago Tiscornia

Page 145 and 146 by Kyle Canan

Page 149, 183, 192, 213, and 223 by Christian Schwarz

Page 160 by Alfred DH

Page 165 and 178 by T. Moult

Page 166 by Jelte Vredenbregt

Page 167 by Benhadja Ahmed Riadh (truth_hunterr)

Page 168 by J.A. Cooper

Page 169 and 170 by David Taylor

Page 169 by Jon Mac Gillivray

Page 170, 171, and 187 by Gerhard Suttner

Page 175 by Jochen Gartz

Page 176, 177, 195, and 208 by Scott Ostuni

Page 179 by Richard Gaines

Page 182 by John W. Allen

Page 186 by Nikola Lačkovič

Page 188 by Giuliani Furci

Page 194 by Jordan T. Jacobs

Page 196 by Beryl Duck

Page 196 by Michael Wallace

Page 197 by Anonymous

Page 198 by D. B. Townsend

Page 199 by James Conway

Page 205 by Julio Alvarez

Page 217 by Tom Bigelow

Page 218 by Mycellenz WikiCommons

Page 219 by Annie Weissman

Page 221 by Werner Diekow

Page 225 by Felice Di Palma-Rome

Page 226 by Taylor Bright

Page 227 by Seymour Burgess

Page 227 by Andrew Kunik

Page 227 by Ivàn Pèrez Muñoz

Index

About the Author

Paul Stamets has been a mycologist for fifty years and is the author of seven books. He has named new species of psilocybin mushrooms, and a new species of mushroom—*Psilocybe stametsii*—has recently been named after him in honor of his lifelong contributions to mycology.

Stamets is an invention ambassador for the American Association for the Advancement of Science and was inducted into The Explorer's Club in 2020. He has received numerous awards, including the National Mycologist Award from the North American Mycological Association, the Gordon and Tina Wasson Award from the Mycological Society of America, the Disruptor Award from NextMed, and the Lifetime Achievement Award from SynBioBeta.

A dedicated Earthling and Knowledge Keeper, Stamets is a leader in the protection of ecosystems focusing on the roles of mycelium and mushrooms and abides by the philosophy that "MycoDiversity is BioSecurity." He funds research to save rare strains of mushrooms that dwell within old-growth forests and believes these forests contain a deep reservoir of species essential for pandemic defense. His contributions include supplying mushroom mycelium to several clinical studies showing immune support. He is a collaborator with numerous academic scientific organizations and research institutes. His research is considered a breakthrough by thought leaders for expanding consciousness that can improve ecosystem health worldwide.

His work entered mainstream popular culture when a character in the Star Trek: Discovery series on CBS, astromycologist Lt. Commander Paul Stamets, was named after him. His work with mycelium has become a central theme of the series. In addition, Stamets is the primary guide to the mushroom documentary *Fantastic Fungi.*

As a wisdom keeper, his life has continued the tradition passing on the knowledge inherited from past generations, making new discoveries to benefit the next generations of mycologists, and building mycological bridges across cultures and continents. Stamets lives in Washington state.

This book aims to provide useful information introducing readers to varieties of mushrooms and their potential uses. Author and publisher are not responsible for any errors in identification of mushroom species and make no representations regarding their safety or health benefits. There is risk to ingesting wild mushrooms; illness and even death can occur. Some mushrooms that are nontoxic for most people may make some people ill. Author and publisher make no warranties as to the safety of consuming wild mushrooms and accept no responsibility for any consequences or health problems resulting from reliance upon information contained herein or that may arise from contact with, or the ingestion of mushrooms described herein. Persons who ingest mushrooms or any other potentially dangerous fungi do so at their own peril and should avoid ingesting mushrooms of unknown origin.

Ten Speed Press
An imprint of the Crown Publishing Group
A division of Penguin Random House LLC
TENSPEED.COM

Typefaces: Lizzie Allen's Blunk and Weltkern's TWK Lausanne

Library of Congress Cataloging-in-Publication Data
Names: Stamets, Paul, author. Title: Psilocybin mushrooms in their natural habitats : a guide to the history, identification, and use of psychoactive fungi / Paul Stamets.
Identifiers: LCCN 2024033051 (print) | LCCN 2024033052 (ebook) | ISBN 9781984863027 (trade paperback) | ISBN 9781984863034 (ebook) Subjects: LCSH: Mushrooms, Hallucinogenic. | Psilocybin.
Classification: LCC QK604.2.H34 S73 2025 (print) | LCC QK604.2.H34 (ebook) | DDC 579.6—dc23/eng/20240829
LC record available at https://lccn.loc.gov/2024033051
LC ebook record available at https://lccn.loc.gov/2024033052

HC ISBN 978-0-593-83900-3
TR ISBN 978-1-9848-6302-7
Ebook ISBN 978-1-9848-6303-4

Editor: Julie Bennett | Production editor: Ashley Pierce
Designer: Lizzie Allen | Production designers: Mari Gill and Faith Hague
Production manager: Philip Leung
Copyeditor: Lisa Brousseau | Proofreaders: Sasha Tropp and Allie Kiekhofer | Indexer: Ken DellaPenta
Publicist: Chloe Aryeh | Marketer: Joey Lozada

Manufactured in China

10 9 8 7 6 5 4 3 2 1

First Edition